D1414726

**Energy
Conservation
for American
Agriculture**

VANIER
SC

Energy Conservation for American Agriculture

Robert A. Friedrich

Environmental Law Institute State and Local Energy
Conservation Project

Ballinger Publishing Company • Cambridge, Massachusetts
A Subsidiary of Harper & Row, Publishers, Inc.

057125

 This book is printed on recycled paper.

This book was prepared with the support of NSF Grant APR 7504814. However, any opinions, findings, conclusions, or recommendations herein are those of the author and do not necessarily reflect the views of NSF.

Copyright © 1978 by Environmental Law Institute. All rights reserved. No part of this publication may be reproduced, stored in a retrieval system, or transmitted in any form or by any means, electronic mechanical photocopy, recording or otherwise, without the prior written consent of the publisher.

International Standard Book Number: 0−88410−058−8

Library of Congress Catalog Card Number: 78−11473

Printed in the United States of America

Library of Congress Cataloging in Publication Data

Friedrich, Robert A
  Energy conservation for American agriculture.

  Includes bibliographical references and index.
  1. Agriculture—United States—Energy conservation.
I. Environmental Law Institute. Energy Conservation
Project. II. Title.
TJ163.5.A37F74   333.7   78−11473
ISBN 0−88410−058−8

TJ
163.5
A37
F74
1978

Dedication

To my mother and father for furnishing the spark.
To Judy for sustaining the fire.

Contents

List of Tables

Preface

In a world where millions go hungry, American agricultural abundance is both a gift and an opportunity. It is a gift to the American people, most of whom seldom fear hunger; it is an opportunity to export food for both money and humanitarian goals. In the recent past, American agriculture has also been seen as a model for the rest of the world. Since the 1973 oil embargo, however, we have come to learn that our plentiful harvests have been bought at a considerable energy price. Our reliance on high quality energy sources, both for day-to-day farm operations and for fertilizing crops, has underlined the close relationship between energy prices and retail food prices. More fundamentally, it has suggested the vulnerability of the nation's food system to disturbances in energy supply.

Agriculture is an American success story. Why should we risk killing the golden goose? Moreover, if we do decide to make changes, how should they be done so that productivity is not imperiled? This book by Robert A. Friedrich addresses these questions. It translates into policy recommendations the findings of recent academic and government reports that quantify potential energy savings in farm machinery operations, irrigation, crop drying, fertilizer and pesticide use, and livestock production.

Although agricultural production uses only a small percentage of our annual energy consumption—currently 3 to 4 percent—significant potential exists for saving part of that energy. Many farmers can economically reduce their energy consumption using current technology. Since farming operations vary from state to state and even from farm to farm, no one governmental strategy can be offered as best. Rather, Friedrich offers a range of ideas for state, county, and city governments to consider, suggesting approaches that can be adapted to local needs. Education and the use of persuasion are themes of this book; regulation and financial penalties, although discussed, play only a small role in the suggested array of strategies. Friedrich believes that assistance, inducements, and information will bring about the necessary changes less disruptively, particularly when those changes are in the economic self-interest of the farmer.

This book is one volume of the Environmental Law Institute's State and Local Energy Conservation Project, a series of books that examines legal and administrative strategies that states and localities can use to promote energy conservation. Funded by a grant from the National Science Foundation, the series offers a range of legal and regulatory approaches to eliminate uneconomic energy waste in various sectors of the economy. Other books in the series deal with land use planning for more energy-efficient cities, government procurement and operations for energy efficiency, industrial energy use, transportation, buildings, governmental financial policies, and utility pricing. The recommendations are based on a study of existing laws and regulations and experience with conservation programs that have already been adopted. The books have been shaped to fit the energy information needs of state and local government officials who are charged with preparing and implementing effective energy conservation programs. But the books are also written to be accessible to citizens and legislators concerned with enhancing energy conservation efforts.

Robert Friedrich's book is a useful and stimulating addition to this series. Energy-efficient agriculture is not to be feared: instead, it is an opportunity to be seized with ingenuity and dedication. Fortunately, these are the traditional character traits of the American farmer.

> **Grant P. Thompson**
> Institute Fellow
> Principal Investigator
> Energy Conservation Project

Acknowledgments

Many persons helped in the preparation of this book. Although it is difficult to choose where to begin, I owe special thanks to my colleagues on the Energy Conservation Project for commenting incisively on draft chapters, for creating a pleasant writing environment, and for providing a strong feeling of camaraderie. Over the many months of the project, their friendship and support helped me through the hard times, and for this, many thanks.

Eric Erdheim, now an attorney at the National Oceanic and Atmospheric Administration, and Ted McConnell, a student at Georgetown University Law Center, provided the principal research assistance. A preliminary draft was reviewed at an early stage by William Lockeretz of the Center for the Biology of Natural Systems and Elinor Cruze Terhune of the National Science Foundation. Their comments were of great assistance in terms both of substance and of structure. As the work progressed, Stephen O. Andersen, a resource economist at the College of the Atlantic, provided keen review and a sense of direction.

The draft manuscript was edited by Page Shepard, Margaret Hough, Gail Boyer Hayes, and Juliet Pierson. Their sharp-penciled criticism helped shape my thoughts into clear prose.

Barbara L. Shaw and Patricia M. Sagurton gathered and organized materials on state energy conservation and provided overall administrative assistance. I discovered the thoroughness of their efforts whenever I needed to use a proposed bill or statute.

Laverne Diggs typed the final draft, maintaining a remarkably high quality throughout a lengthy and difficult manuscript. Earlier drafts were ably typed by Janet Coffin, Ann Falk, Pat Hayden, Karla Heimann, Sherry Leichman, Tina Luebke, and Stephanie Nagata.

A final word of thanks goes to Grant P. Thompson, director of the Energy Conservation Project, for offering me those devilishly hot jalapeño peppers in the first place.

Chapter 1

Energy and Agriculture

Man has, through the ages, evolved a strategy for manipulating the plant and its environment to maximize solar energy conversion into food, feed, and fiber. It is called agriculture. . . . The challenge now is to reduce energy inputs into food production systems without jeopardizing productivity or energy output.[1]

Energy and agriculture. The very words bring to mind an interwoven image of farmers as both energy consumers and energy producers. This image is accurate, for farm operations each year consume billions of gallons of refined petroleum products and millions of tons of petroleum-based fertilizers and pesticides. At the same time, farming is an energy-producing sector, as solar energy is harnessed by plant photosynthesis to produce the bread we eat, the milk we drink, and the cotton and wool we wear. Remembering that the modern American farmer is a sophisticated specialist responsible for producing a plentiful harvest and for conserving scarce energy resources will help place in perspective the techniques and strategies discussed throughout the book.

PURPOSE AND SCOPE OF BOOK

This book looks at energy use on farms and attempts to identify energy-efficient strategies that states and localities should promote. Many barriers to energy conservation are created by institutional arrangements, by economic misallocations, and by legislation that invites energy waste. These are the barriers that will be scrutinized. The book is addressed to state and local officials, farmers and farm organizations, and others who want to better understand how energy is used in agriculture and how it can be conserved.

Throughout the book, ideas are presented to show how farmers could make better use of energy. Some ideas are simply good housekeeping changes, drawing on existing technology and expertise to bring quick but small energy savings. Other ideas describe the current best practices, the so-called state of the art. These ideas, while still utilizing current technology, require some investment to modify equipment for improved energy efficiency and will generally yield energy- and cost-savings over a longer period of time. A third category of changes requires farmers to convert to alternative processes that promise improved energy efficiency and long-term cost reduction, but entail a substantial initial capital investment. The book shies away from offering suggestions that go beyond this third level. Changes to novel processes or entirely new production methods may save energy but they entail economic risks and major conversions that are more properly the province of national agricultural policymakers.

CONTENTS OF BOOK

Chapter 2 examines the range of options available to policymakers to encourage or mandate energy conservation. Financial incentive programs, taxes, regulatory policies, and information transfer programs are described and analyzed. Chapter 3 considers how energy is directly used on farms for farm machinery operations, irrigation, crop drying, and specialized activities. New approaches for increasing the energy efficiency of these operations are described and assessed. Chapter 4 examines the indirect use of energy on farms and analyzes techniques and strategies for reducing the use of energy-intensive fertilizers and pesticides. Chapter 5 focuses on livestock operations, a particularly energy-intensive segment of agricultural production. The production of beef and other livestock is frequently criticized as energy-inefficient, and this chapter considers the practicality and energy-conserving potential of some suggested changes. Chapter 6 briefly examines food production in an urban environment. Gardening and livestock raising are increasingly popular activities that can contribute to energy savings in the food system. The book concludes by stressing the importance of government leadership in devising agricultural policies to help conserve energy.

HOW MUCH ENERGY DOES AGRICULTURE USE?

Before embarking on an examination of energy use and energy conservation, it is important to have a common understanding of what is meant by the term "energy." Farmers who work a full day inside the cab of a large combine have certainly exerted energy—human energy. Likewise, the corn and soybean plants that photosynthesize sunlight embody another form of energy—solar energy. Neither human labor nor sunlight, however, is considered here in calculating energy consumption levels or energy conservation potential. Rather, the focus is on fossil fuels, nonrenewable resources that farmers use in the form of petroleum fuels, electricity, fertilizers, and synthetic pesticides.

If the history of agricultural development in the Western world is viewed as an evolution of energy use, the 9,000 years of farming can be divided into three specific ages. The first age, which lasted until the end of the nineteenth century, found farmers cultivating the land, using their own muscle power and that of draft animals. In this way, farmers were able to capture the sun's energy to expand food production beyond what was available when people were simple gatherers and hunters.

A transition to mechanized farming occurred during the second age. Energy traditionally supplied by physical labor was supplemented by new farm machinery and fossil fuel energy. Compared to the first age, this period was but a brief moment, lasting from the late nineteenth century to the middle of the twentieth.

The third age is a scant three decades old, yet the changes that have occurred in these few years have been phenomenal. Since the 1940s, American agriculture has been based almost entirely on fossil fuel energy, and today the production of food and fiber consumes enormous quantities of refined petroleum fuels, natural gas, and electricity to power tractors, irrigate fields, and fertilize crops. This energy subsidy has yielded astonishing results. At a time when much of the world still continues to struggle with the problem of producing enough food, American farmers have had extraordinary success in meeting the needs of the nation and producing surpluses to help feed a hungry world. Between 1920 and 1940 farm output per acre remained relatively unchanged, but since 1940 the output on U.S. farms has increased at an annual rate of almost 2 percent.[2]

Despite these impressive gains, the plentiful harvest has been bought at a considerable cost: modern American agriculture is almost totally dependent on the availability of fossil fuel energy. This dependence has far-reaching implications. First, American agriculture and the entire food production system are increasingly vulnerable to disturbances. If the Arab oil embargo in 1973–1974 did not directly affect agricultural output, it did show policymakers the problems that could arise if the nation's supply of imported oil were suddenly cut off. Second, reliance on refined petroleum products, electricity, and natural gas also contributes to economic uncertainty. The combined changes that have occurred in the wake of the oil embargo created ripples that have spread out to touch the farmer, who faces higher costs for fuels, fertilizers, and pesticides, and to the consumer, who finds these costs passed on as higher food prices at the supermarket.

Recent studies have calculated that American farmers now expend approximately 2 to 3 quadrillion BTUs, or roughly 3 to 4 percent of all energy used in the nation.[3] While large in an absolute sense, the amount of energy used is but a tiny percentage of all energy consumption. Even if one considers the amount of energy used to bring food from sown seed to the consumer's stomach, the entire food system—including production, processing, marketing, and preparation—accounts for only 12 to 17 percent of the total national energy budget.[4] Thus, compared to the amount of energy consumed in industrial processes (37 percent of total energy use) or transportation (26 percent), energy consumption in agriculture is truly small potatoes.[5]

WHY FOCUS ON ENERGY CONSERVATION?

If agriculture uses only a small percentage of the fossil fuel energy consumed in this country, a natural question is, Why spend time, money, and effort in an attempt to save this tiny fraction of the national energy budget? The short answer is that a significant potential exists for increasing energy efficiency in agriculture: trillions of BTUs, representing a healthy percentage of current use, can be saved economically.

Few people would seriously suggest that farmers discontinue their use of nonrenewable fossil fuels, but many critics do urge a reconsideration of the way energy is used. There are many situations in which modern farmers spend more fossil fuel energy than they recover in the form of food. The production of most meats and dairy products and many fruits and vegetables falls into this category. Of course, this observation must be tempered by one consideration: unlike petroleum, which cannot be eaten, these foods yield high quality proteins, fats, and carbohydrates. Still, these foods require more energy to produce than they yield, and policymakers would do well to encourage the production of these goods in a more energy-efficient manner.

Some crops, such as corn and wheat, yield more food energy than is consumed in their production. Several studies, however, have concluded that the energy efficiency with which they are now produced has declined significantly in recent years.[6] Despite dramatically increased crop yields, energy consumption has grown at an even faster pace. Many experts are troubled by this, and policymakers must grapple with the problem of stemming or reversing the trend.

Researchers and analysts both in and out of government see great promise for energy-conserving strategies that are easily implemented and that maintain crop yields. For example, in 1975 a spokesman for the Federal Energy Administration (FEA) estimated that about 500 trillion BTUs could be saved annually by 1985, a significant 20 percent reduction in the current 2.5 quadrillion BTUs now consumed.[7] Even more encouraging is a report prepared for the U.S. Energy Research and Development Administration (ERDA) that estimates a realistic potential saving of 700 trillion BTUs in food production.[8] Researchers who have studied specific crops, such as corn, or specific techniques, such as the use of irrigation, report potential energy savings of up to 50 percent without any decrease in crop yield.[9]

Although the public is constantly bombarded with ingenious ideas to produce energy, it remains likely that energy production costs will continue to climb steadily for the rest of the century. The cold fact is that, for the next quarter century, the best way to produce more energy is to save it.

The federal government seems to acknowledge that conclusion in its assessment of energy conservation as an integral part of our national energy policy. In its 1976 annual report to Congress, ERDA described the benefits of energy conservation as follows:

- A barrel of oil saved can result in reduced imports.

- It typically costs less to save a barrel of oil than to produce one through the development of new technology.

- Energy conservation generally has a beneficial effect on the environment in comparison to energy produced and used.

- Capital requirements to increase energy use efficiency are generally lower than capital needs to produce an equivalent amount of energy from new sources since most new supply technologies are highly capital intensive.

- Conservation technologies can generally be implemented at a faster rate and with less government involvement in the near term than can new supply technologies.

- Energy efficiency actions can reduce the pressure for accelerated introduction of new supply technologies. Since the actions persist over time, the benefits are continuing.[10]

Energy conservation can slow the growth rate in total energy demand, thereby softening the economic impact of higher energy prices. Energy conservation also gives decisionmakers more time in which to find an acceptable long-term solution to our energy supply problem.

Unfortunately, energy conservation is often mistakenly depicted by energy companies and even by politicians to mean sacrifice and curtailment of human activities. It does not. It means more effective use of energy and other resources—in other words, doing better, not doing without. In many instances, farmers can benefit from adopting energy-conserving measures. Investment in energy conservation is frequently cost-effective and helps hold the line against increased production costs. For many farmers, raising their energy awareness is just a matter of pointing out the economic thing to do. Like cost consciousness, energy awareness is built into any successful farm operation.

OBSTACLES TO ENERGY CONSERVATION

Despite the abundant potential for energy conservation in agriculture, states and localities should realize that obstacles do exist. It is likely that one or more of these impediments will limit the true potential, and decisionmakers should not be surprised to find the road to energy conservation lined with hurdles.

First, competing policies exist at the federal level regarding agricultural production and energy conservation, making the full achievement of each goal improbable. Our current national agricultural policy encourages farmers to produce as much as possible. All-out production, however, usually requires ever-increasing amounts of energy input for each additional unit of output. Simultaneously, our national energy policy seeks to bring energy demand and energy supply into balance by decreasing demand and increasing supply. As part of this policy, energy conservation is accorded the highest priority, and farmers, like everyone else, are expected to use fossil fuels ever more effectively in producing farm commodities. States should expect that the tension created by these competing national policies is likely to result in neither maximum food production nor maximum energy conservation, but in some compromise between these goals.

Second, the federal government has a commitment to maintain reasonable food prices. This policy might limit the implementation of some strategies that could effectively conserve energy but could also lead to higher retail food prices. Price-inflating strategies are politically unfeasible and techniques that tend to boost food prices are unlikely to be adopted, irrespective of their energy-conserving merits.

Third, it is the responsibility of any government to pursue policies that will guarantee an uninterrupted supply of food. Energy conservation is not an end in itself, and few policymakers would be so foolish as to risk food shortages to save energy. Farming entails risks, and both farmers and agricultural planners must take into account such risks as unfavorable weather or unusual pest infestations. To protect against severe food losses, farmers have used and will continue to use certain quantities of fuel and energy-intensive pesticides over and above minimum requirements. These energy "wastes" may have to be accepted as the insurance premium paid to protect against food shortages.

Other federal policies may limit what states can do. For example, federal and state environmental regulations and programs may prohibit some energy-conserving techniques. Energy pricing policy, too, might interfere with energy conservation. Certainly this is true of the artificially low price of federally regulated interstate natural gas, whose low cost tends to result in its misallocation.

Finally, a state's finite resources of time, funds, and personnel limit the development and implementation of energy conservation policies and programs. As a practical matter, energy conservation is but one goal that must compete for resources with other state programs, including other energy programs.

LIMITS TO THIS BOOK

This book does not cover energy conservation in the entire food system, but concentrates specifically on agricultural production. Although this limitation precludes consideration of a substantial part of the food chain where opportunities exist for significant energy savings, the focus of this book would be blurred if it attempted to cover food processing, transportation, marketing, and preparation. These additional links in the food chain are separable from farm production in many respects: different energy-conserving techniques are used, different constraints must be overcome to achieve energy savings, and different government agencies are responsible for policy planning. Happily, some energy conservation strategies that can be used in these parts of the food system are covered in other books of this series.[11]

The strategies recommended in this book are far from comprehensive. Neither are they equally effective, economical, or easy to use. Rather than attempt to set out one "best" solution, this book offers many strategies that are varied, sometimes contradictory, and rarely to everyone's liking. This choice is made consciously to stimulate discussion and to inform policymakers of the often competing considerations in selecting a particular strategy. By being presented with a full range of options, state agricultural policy planners can pick and choose the implementation strategies that best meet the needs of the farmer and the needs of the state.

NOTES TO CHAPTER 1

1. S.H. Wittwer, "Food Production: Technology and the Resource Base," *Science* 188 (May 9, 1975): 582. Copyright © 1975 by the American Association for the Advancement of Science. Used with the permission of the author and the copyright holder.

2. U.S. Department of Agriculture, Economic Research Service, *Changes in Farm Production and Efficiency*, Statistical Bulletin no. 561 (Washington, D.C.: U.S. Government Printing Office, 1976), p. 68.

3. To simplify energy comparisons, the common energy unit used throughout this book is the British Thermal Unit (BTU). One BTU is the amount of energy required to heat one pound of water from 62°F to 63°F. The kilocalorie (kcal), frequently used in research reports, is equal to 3.969 BTUs.

Modern farming operations use gasoline, diesel fuel, fuel oil, liquefied petroleum (LP) gas, natural gas, and electricity. The conversion factors used by the U.S. Department of Agriculture are:

Gasoline	124,000 BTUs per gallon
Diesel Fuel	140,000 BTUs per gallon
Fuel Oil	138,000 BTUs per gallon
LP Gas	92,000 BTUs per gallon
Natural Gas	1,068 BTUs per cubic foot
Electricity	3,412 BTUs per kilowatt-hour

The estimate of total energy consumption in agricultural production is taken from several sources. *See, e.g.*, Allen Schienbein, *A Guide to Energy Savings for the Field Crops Producer*, prepared for the U.S. Department of Agriculture and the Federal Energy Administration (Washington, D.C.: U.S. Department of Agriculture, 1977), pp. 1–3; Booz, Allen & Hamilton, Inc., *Energy Use in the Food System*, prepared for the Federal Energy Administration (Washington, D.C.: U.S. Government Printing Office, 1976), pp. IV–6 to IV–9; U.S. Department of

Agriculture, Economic Research Service, *The U.S. Food and Fiber Sector: Energy Use and Outlook*, prepared for the U.S. Senate Committee on Agriculture and Forestry (Washington, D.C.: U.S. Government Printing Office, 1974), pp. v-xv; John S. Steinhart and Carol E. Steinhart, "Energy Use in the U.S. Food System," *Science* 184 (April 19, 1974): 309; and Eric Hirst, "Food-Related Energy Requirements," *Science* 184 (April 12, 1974): 136.

4. Hirst, *supra* note 3 at 134–38; Booz, Allen & Hamilton, Inc., *supra* note 3 at IV–18 to IV–25.

5. Federal Energy Administration, *Monthly Energy Review* (Springfield, Virginia: U.S. Department of Commerce, National Technical Information Service, January 1977), pp. 50–55.

6. *See, e.g.*, David Pimentel et al., "Food Production and the Energy Crisis," *Science* 182 (November 2, 1973): 446.

7. "Conservation: FEA looks at Agriculture for Production Energy Conservation Potential," *BNA Energy Users Rep.* no. 97 (June 19, 1975): A–6.

8. John D. Buffington and Jerrold H. Zar, "Realistic and Unrealistic Energy Conservation Potential in Agriculture," in *Agriculture and Energy*, ed. William Lockeretz (New York: Academic Press, 1977), pp. 695, 708.

9. Pimentel et al., *supra* note 6 at 447; D.E. Lane, P.E. Fischbach, and N.C. Teter, *Energy Uses in Nebraska Agriculture* (Lincoln: University of Nebraska, College of Agriculture and Home Economics Extension Service, 1973), p. 17.

10. *See* U.S. Energy Research and Development Administration, *A National Plan for Energy Research, Development & Demonstration*, ERDA 76–1, 2 vols. (Washington, D.C.: U.S. Government Printing Office, 1976), 1:8.

11. *See* Norman L. Dean, *Energy Efficiency in Industry* (Cambridge, Massachusetts: Ballinger Publishing Co., forthcoming); Grant P. Thompson, *Building to Save Energy—Legal and Regulatory Approaches* (Cambridge, Massachusetts: Ballinger Publishing Co., forthcoming).

✳ *Chapter 2*

Options for Change

On acknowledging the full costs of energy waste and the potential benefits of energy conservation, policymakers must proceed from the question of whether to encourage conservation to the harder question of how to do it. This chapter tells about the tools available to states planning comprehensive energy conservation programs. These policy options fall into four broad categories: financial incentives, taxes, regulations, and information transfer programs.

The first category, financial incentives, can effectively reduce the costs and risks of investment in energy-conserving equipment and techniques, thereby persuading farmers to alter present patterns of energy consumption. State and local governments can encourage such investments through spending programs that fashion official largess into cash grants or loan programs, or they can offer tax incentives to induce energy-conserving investments. By offering other subsidies, such as the sale of energy-saving equipment at discount or the use of municipal land for public gardens, further energy savings can result.

Financial disincentives, including taxes on fuels or on electricity, can save energy by discouraging the waste of natural resources. Imposing such taxes on farmers, however, will tend to raise production costs as well as to discourage energy waste. Since increased food production costs usually mean lower farm profits or higher retail food prices or both, farmer and consumer opposition to such measures is sure to be strong.

A third tack to modifying energy consumption patterns is direct state regulation. State agencies can, for example, set quotas on consumption, prohibit certain uses of fuel, and control the price and the availability of electricity. Like financial disincentives, regulations must be implemented with care to keep food production up and retail food prices down to a reasonable level.

Finally, new ideas can help to improve energy efficiency on farms. Increased state and local funding of research, development, demonstration projects, and educational programs can inform farmers of the latest agricultural technology. With the aid of their local agricultural extension specialists or of fuel efficiency labeling on farm equipment, farmers can become more aware of and better able to practice energy conservation.

Of the four policy options outlined above, financial incentives and information transfer programs will encourage energy conservation in a positive, noncoercive way unlikely to provoke opposition from most farmers. Financial disincentives and regulatory schemes are more coercive approaches that may inspire farmer evasion or heel-dragging compliance, and care must be taken to devise taxes or regulations that will not penalize farmers. In sum, policymakers devising programs to reduce energy use and increase energy efficiency in agriculture must proceed cautiously, since they are dealing with a basic life necessity.

Because the concerns of different states or regions vary, no "best" approach is presented; rather, several strategies are described and analyzed. Policymakers must weigh the costs of a strategy (including the financial investment and expected adverse effects on food production and retail prices) against its benefits (energy conserved and environment preserved). Where the energy conservation impact is adequate, noncoercive credit subsidies, tax incentives, and information transfer programs are preferable to taxation and regulation, for several reasons:

- Politically, they will be more popular;
- Administratively, they will cost less;
- Legally, they will face fewer challenges;
- Practically, they will be less apt to inhibit food production or to raise food prices.

FINANCIAL INCENTIVES

Efforts to conserve energy in agriculture are stymied by at least two significant financial limitations. First, many farmers wishing to buy new, more energy-efficient equipment or to retrofit the energy-inefficient equipment they already own cannot obtain sufficient financing at an attractive interest rate. Second, other farmers lack even the motivation to invest in new equipment or improvements; the potential savings look too small to justify the economic risk. Through spending programs and tax incentives that benefit farmers who conserve energy, governments can relieve these financial constraints.

Of all government powers, the power to spend money is probably the most politically acceptable; it infringes on no individual rights and forces no one to obey against his or her will. State and local governments can encourage investment in energy-conserving equipment by offering farmers direct cash grants, low interest loans, purchase subsidies, and benefits in kind. Of course, all these spending programs call for the allocation of large sums of money, and state and local governments must keep a tight rein on program disbursements to stretch the energy conservation effect of every dollar. This message seems especially significant now, in light of the overwhelming voter approval of the Jarvis-Gann initiative in California, amending the constitution to limit local property taxes in that state. The resulting aftershock has reached polic 'makers at all levels of government throughout the country, and it ɜems safe to say that the so-called "taxpayer's revolt" in California will affect spending programs in other states as well.

The well-accepted general rule for the spending of public funds is that a state legislature may appropriate and expend money for any purpose as long as its action violates no limitation, either explicit or implied, appearing in the state or federal constitutions. With energy use impinging upon nearly every aspect of life within a state, including public health, employment and the economy, and the environment, both energy waste and energy conservation are valid concerns of state legislators. An expenditure of state funds in the forms of cash grants and subsidized loans thus falls well within state powers to spend for the general welfare.

Along with a state's power to spend for the general welfare is its authority to devise tax programs. State discretion is broad as long as those programs comply with the U.S. Constitution, state constitution, and statutes. Local governments, too, may set taxes when that authority has been delegated by the state. In addition to raising revenue, tax programs can influence an individual farmer's energy conservation decisions, and many states have already decided to stimulate investment in energy-conserving equipment by offering tax exemptions, deductions, and credits.

Enacting tax subsidies and loan programs to encourage investment in energy conservation has proven to be a popular legislative response to the "energy crisis." There are several readily apparent advantages that may explain this response. An offer of financial incentives will, by removing some economic barriers, convince at least some farmers to modify their energy use decisions. The marginal effect will, of course, depend on the magnitude of the incentive. For legislators, financial incentives have the virtue of provoking little, if any, opposition. Few people consider government subsidies novel or shocking, and today the recipients of government welfare expenditures range from individual college students with state-guaranteed low interest loans to large industries protected by stiff import tariffs. Although government intervention in the marketplace raises important philosophical questions, state and federal policies have frequently considered particular social goals sufficiently vital to warrant such interference, even when the benefits and burdens do not fall equally on all income classes. Energy conservation could well join the list of social goals worthy of government subsidies, and financial incentives for farmers to save energy can be viewed as an investment in an orderly transition to an era of renewable energy sources.

Nevertheless, strong arguments can be made against offering financial incentives.[1] Perhaps the most convincing argument is that it is almost impossible to create a practical, workable subsidy program. If the aim of a state energy conservation program is to encourage investments that would not occur in the absence of the incentive, the definition of an eligible investment must be broad enough to encourage investment in energy conservation, yet narrow enough to avoid windfalls to those who would have invested anyway. Drafting such a definition will tax the most creative legislative draftsmen. These drafting difficulties are illustrated in two state energy conservation incentive statutes, one enacted and the other proposed.

Montana grants preferential tax status to investments made for an energy conservation purpose, which is defined as "one or more of the following results of an investment: reducing the waste or dissipation of energy, or reducing the amount of energy required to accomplish a given quantity of work."[2] This definition may be overbroad in its reach, since a clever farmer might use it to claim almost any business expenditure as an investment that is eligible for preferential tax treatment.

At the other extreme, a proposed Maryland bill providing income tax credits defines qualified energy conservation investments as

> improvements, materials and equipment designed and installed in a building or on land for the utilization of solar heat, wind, solid wastes, or the

decomposition of organic wastes, for capturing energy or converting energy sources into useable sources, or for the production of electric power from solid wood wastes, or a small system for the utilization of water power by means of an impoundment of not over twenty acres in surface area.[3]

This approach narrows the range of eligible investments, but it does so at the cost of excluding some significant conservation investments, including the installation of insulation or the use of some newly discovered energy-saving device.

A better attempt at defining energy conservation investments eligible for an incentive is found in the federal Energy Conservation and Production Act of 1976, which defines an energy conservation measure as one

which modifies any building or industrial plant, the construction of which has been completed prior to August 16, 1976, if such measure has been determined by means of an energy audit or by the Administrator, by rule under section 6325(e)(1) of this title, to be likely to improve the efficiency of energy use and to reduce energy costs (as calculated on the basis of energy costs reasonably projected over time, as determined by the Administrator) in an amount sufficient to enable a person to recover the total cost of purchasing and installing such measure (without regard to any tax benefit or Federal financial assistance applicable thereto) within the period of—

(A) the useful life of the modification involved, as determined by the Administrator, or

(B) 15 years after the purchase and installation of such measure, whichever is less. Such term does not include (i) the purchase or installation of any appliance, (ii) any conversion from one fuel or source of energy to another which is of a type which the Administrator, by rule, determines is ineligible on the basis that such type of conversion is inconsistent with national policy with respect to energy conservation or reduction of imports of fuels, or (iii) any measure, or type of measure, which the Administrator determines does not have as its primary purpose an improvement in efficiency of energy use.[4]

While this is not an entirely satisfactory definition, it represents a vast improvement over those attempted at the state level and, on the whole, provides a useful model for states.

In addition to the problem of defining eligible investments, using financial incentives to encourage energy conservation has other shortcomings. Many economists argue that subsidies tend to distort the market by altering the cost of one investment relative to another, often with unforeseen results. Incentive programs frequently outlive their original purpose, persisting long after their usefulness has vanished. Tax incentives have been further criticized for hiding the true cost of these subsidies from both taxpayers and policymakers and for increasing the complexity of the tax system.

In sum, states should cautiously explore the use of financial incentives in their search for ways to encourage energy conservation on farms. The recent spate of state energy conservation legislation confirms that financial incentives, particularly tax incentives, have strong support. This interest seems likely to continue, and the remainder of this section discusses legislative strategies that states can consider, legal problems that may arise, and methods of avoiding or minimizing these problems.

Policymakers who decide to encourage farmers to invest large sums in energy-conserving new equipment or to retrofit old equipment may choose from a wide variety of incentive programs. In particular, states can consider using cash grants, loan programs, the purchase and sale of farm equipment at discount, and tax allowances.

Cash Grants

Cash payments to farmers could take the form of construction grants, capital equipment grants, or operating grants. Examples of this sort of financial incentive abound at the federal level, among them operating subsidies the Civil Aeronautics Board grants commercial air carriers and construction subsidies from the Federal Maritime Administration to American shipbuilders.

Direct cash subsidies get results; they are straightforward and their message is clear. Their high degree of visibility ensures the attention of potential recipients and of those involved in budgeting and appropriating funds. Since the funds for cash grants must be appropriated by legislators, this technique offers policymakers great flexibility to increase or decrease funding as conditions change. Cash grants allow decisionmakers to subsidize farmers who use energy-efficient, labor-intensive techniques as well as those who invest in energy-efficient capital equipment. And the benefits of cash incentives accrue equally to all who receive them, unlike tax incentives whose benefits go mainly to large or profitable farming operations.

Certain other aspects of the cash payment approach should give prudent policymakers pause. First, only large sums of money can effect significant energy conservation gains under a direct cash program, and one must ask, Where will the money come from? Second, such programs are bound to meet with stiff political opposition from those who are *not* to receive any cash. Third, the administrative complexity of cash subsidy programs will mean unwelcome additional expense and a tangle of red tape. Finally, cash subsidies may result in windfalls to many farmers who will be paid for doing exactly what they would have done in the absence of these programs.

Perhaps states could limit the application of cash inventives by offering them only to farmers who invest in new equipment that promises substantial energy savings but poses unusual economic risks. For example, states might subsidize investment in solar equipment to heat or cool farm buildings or in waste heat recovery systems to heat greenhouses. In an effort to stimulate the development of renewable natural resources, Montana has recently enacted laws permitting renewable resource development grants to be made to any department, agency, board, commission, or other division of state government and to individuals.[5] Other states could follow this example to provide seed money for energy conservation research by the state land-grant colleges, by state agricultural researchers, or by individual farmers.

Loan Programs

Before farmers can invest in energy-conserving equipment, they must find a way to finance the investment at reasonable cost. Most banks and other traditional lending institutions will lend to farmers, but will charge interest rates that could wipe out savings expected as a result of the improvement. States can help fill in this gap in the capital market by offering credit subsidies.

A borrower's ability to obtain necessary capital at an interest rate that will leave some room for profit depends upon his or her status in the eyes of the lender. Private lending institutions are often reluctant to lend to people with adventurous investment plans because of increased risk. Through credit subsidy programs, states can reduce the cost of borrowing and minimize default, to the benefit of both farmers and bankers. To increase credit availability, states can set an example by lending state funds directly at low interest rates and by offering loan guarantees.

Direct State Loans. A state could directly loan money to farmers to buy energy-conserving equipment. Such a program could take the form of a Montana law, which authorizes the state to make loans to farmers and ranchers for the conservation and development of renewable resources.[6]

The use of direct state loans at low interest rates offers several advantages. Such loans give the state the ability to direct money into high priority programs with significant energy-conserving potential. A well-designed state loan program can be structured to make investment economically advantageous *and* to require recipients to share in the cost of investment. State loans also share the best qualities of cash grants: eligibility can be made explicit, disbursements can be made accountable, and appropriated budgets can be revised if needs change. In addition, direct state loans have the advantage of not distorting the functioning of the marketplace as much as the use of cash grants or tax allowances.

Before implementing such a program, states should consider some of the drawbacks to direct state loans. Although potentially effective, direct loans have certain costs, including administrative expenses, the cost of loans defaulted on, and the difference between the interest received from borrowers and the interest the state must pay to borrow money. Like tax incentives, the use of state loans would be limited to capital-intensive projects, creating a bias against investment in energy-efficient operations that are labor-intensive. Finally, government entry into the lending business can spell political trouble, as government loans encroach on the business of financial institutions.

To insure maximum effectiveness in the use of loan funds, state officials can set limitations to include:

- Restricting loan eligibility to those investments deemed most likely to result in a significant increase in the efficiency of energy use or conversion,
- Restricting loan eligibility on the basis of financial need,
- Placing a ceiling on the dollar amount of any loan, and
- Limiting the percentage participation by the state on a project.

In addition, a state might wish to empower commercial lending institutions to administer the program for a fee, in order to avoid the creation of a costly administrative bureaucracy and to sidestep political opposition from the financial community.

Implementing a program of direct state loans may require special treatment. Some states will be able to establish a loan fund by a simple act of the legislature. Where general obligation bonds are to be used as the source of loan funds, however, voter approval is required as well as legislative action. Some state constitutions prohibit the extension of state credit for the use of a private person; in such states, a constitutional amendment may be an additional hurdle to the establishment of a conservation loan fund.

Loan Guarantees. A state could guarantee loans to farmers who purchase equipment that met defined energy conservation standards. Under such a program, farmers would borrow money from a traditional lending source, and the state would guarantee repayment of the principal and interest. State loan guarantees reduce the risk of loss to lending institutions and encourage them to lend to certain borrowers at a low interest rate. The idea of loan guarantees is not new. The federal government guarantees some loans administered by the U.S. Small Business Administration and by the Export-Import Bank, but a review of recent state legislation found no state offering this subsidy.

The loan guarantee is a relatively inexpensive device for states to adopt; the only costs are for administration and for picking up the small percentage of loan defaults. Like direct state loans, loan guarantees are accountable and flexible to change; they ultimately place the costs of the guaranteed project on the borrower. Both lenders and borrowers benefit from a loan guarantee program: borrowers receive capital at below market interest rates and lenders make a profitable loan with little risk of financial loss. Loan guarantees allow states to complement, rather than compete with, private lending institutions.

On the other hand, several questions have been raised about the value of loan guarantee programs. Some critics question whether loan guarantees actually serve to increase investment. Often, loan guarantee program administrators set loan evaluation criteria as rigid as those used in the private money market. This approach may reduce loan defaults, but it also reduces the number of loans and the amount of investment. Critics have also questioned whether loan guarantees encourage defaults and point to the experience under the student loan guarantee program with its unexpectedly high percentage of defaults. Finally, loan guarantee programs have sometimes proven unworkable for lack of cooperation from private lending institutions, which may find the "red tape" too burdensome.

A recent proposal considered by Michigan legislators illustrates a possible format for a state loan guarantee program. The Michigan bill would establish a state economic development corporation with a special division for energy research and resource development. One of the duties of that division would be to promote and finance conservation of the state's energy resources through loan guarantees. The state would appropriate money to the economic development corporation, which in turn would assess borrowers 1 percent of the amount guaranteed to recover expenses. In order to insure the stability and viability of the loan guarantee fund, loan guarantees would be subject to the following restrictions:

- Any one guarantee must not exceed 90 percent of the total principal amount of the loan.

- The amount of the guarantee must not exceed 5 percent of the total amount then in the loan guarantee fund.

- The borrower must pay at least 15 percent of the total project costs from nonborrowed funds.

- Making the guarantee must result in a significant cost reduction to the borrower.

- The amount of money in the loan guarantee fund must never fall below an amount equal to the total dollar amount of all outstanding loan guarantees plus the total dollar amount of the guarantee to be made.

- The amount of the guarantee shall not exceed the limit specified in the U.S. Internal Revenue Code.

- The guarantee shall be made only upon application of the financial institution making the loan to the borrower.

- The guarantee shall be in writing.[7]

Naturally, implementing a loan guarantee program in a particular state will require individual tailoring to that state's laws and traditions. In some states, agencies with the power to guarantee loans to farmers may already exist; state officials need only redirect state funds to energy conservation. In other states, no agency exists, and a loan guarantee program can be started by a simple act of the legislature. In still other states, a simple legislative act may be insufficient if the state constitution has a clause forbidding the extension of state credit to private persons; here, a constitutional amendment may be necessary before a loan guarantee program for energy conservation can be begun.

Purchase Subsidies

Some would-be manufacturers and potential purchasers of innovative, energy-efficient equipment fear the specter of financial ruin. States can exorcise this ghost by using their power to purchase and sell goods at prices that encourage production and consumption of those goods.

To reduce the financial risks for manufacturers and purchasers, states can subsidize them both. For farmers who are reluctant to pay a high initial cost for an unconventional but socially beneficial item, states could buy in quantity at the market price and resell at a discount. In this way, the risks to manufacturers and farmers are significantly diminished: the manufacturer is guaranteed a sale at a fixed price and the farmer is able to purchase energy-saving equipment at a discount.

A number of innovative practices hold great promise for energy conservation—supplementation of chemical nitrogen fertilizer with organic wastes, reductions in the use of chemical pesticides, and new crop drying techniques are only three—but these also harbor high risks of crop yield reduction. To protect the economic interests of the farmers willing to try such new production techniques, states could offer crop protection insurance with subsidized premiums. The federal government presently has a crop insurance program, and state-subsidized crop insurance programs could supplement this.[8]

The chief advantage of purchase subsidies is their direct encouragement of both the manufacturer and the user of energy-efficient equipment. But this simplicity at the recipient end is paid for in expense and complexity by the donor, the state. These programs could cost a lot of money; to economize on such programs is to trade off conservation of energy against balance in the budget. Having state governments enter into the distribution system may be objectionable on grounds of interference with the play of market forces. Finally, subsidies of this sort are complex and costly to administer, as illustrated by the history of U.S. Department of Agriculture farm price support programs.

Tax Incentives

The most frequently suggested proposal for encouraging energy conservation on farms is to give farmers a tax break. Many state legislatures have considered some type of tax incentive to reduce the effective cost of investment in energy conservation. Although tax incentives are an option that policymakers will want to consider, they should carefully weigh the competing policy arguments and practical difficulties.

In general, suggested tax incentives to foster energy conservation have taken four forms: investment tax credits, accelerated depreciation, property tax exemptions, and exemptions from sales and use taxes.

Investment Tax Credits. An investment tax credit permits taxpayers to subtract a percentage of the cost of capital equipment from the state income tax owed for the taxable year. For example, a 10 percent investment tax credit would enable a cattle rancher purchasing a $40,000 methane digester to take $4,000 off the amount of income tax owed to the state. Some states have acted to permit excess tax credits to be carried forward to future tax years or to be refunded, a step designed to equalize the tax benefits available to all eligible taxpayers. Allowing investment tax credits reduces the effective cost of purchasing energy-efficient equipment and, if properly designed and publicized, will lead to more investment in energy-conserving items.

Several states have enacted laws permitting investment tax credits for the purchase of energy-saving equipment. Farmers in New Mexico now enjoy an income tax credit of up to $25,000 for the purchase of equipment to construct a solar irrigation pumping system.[9] Kansas encourages all businesses to invest in solar and wind energy systems by allowing a tax credit of 25 percent of the cost, up to a maximum of $3,000.[10] And Minnesota has considered, but not passed, legislation that would give tax credits to those who invested in alternative energy sources, including solar energy systems, wind systems, and the generation of fuel or electricity from agricultural wastes.[11]

Accelerated Depreciation.   Farmers, like other businessmen who purchase capital equipment for the production of income, are permitted to deduct from taxable income an amount attributable to the value of the equipment lost annually because of wear and age. Typically, a farmer uses a standard depreciation table to calculate the percentage that may be depreciated for tax purposes. For example, suppose a farmer purchases an energy-efficient heat recovery system costing $30,000, having an estimated useful life of ten years and no salvage value at the end of that time. If a state allows the farmer to depreciate his equipment using the simple straight line method, 10 percent, or $3,000, may be deducted from taxable income for each of the ten years. If, however, the state permits the farmer to shorten the time period over which an item may be depreciated, the value of the deduction is greater because the deductions are larger and their benefit more immediate. Turning again to the example of the heat recovery system, if the farmer were permitted to depreciate the system over a period of, say, five years rather than its ten year life span, the deduction would be $6,000 for each of the next five years.

Arizona farmers who purchase solar equipment are permitted to amortize the cost over a three-year period, resulting in tax deductions larger than those allowed for depreciation over the actual expected life span.[12] Kansas and Texas also allow solar equipment to be amortized rapidly, although in both states the period is five years and in Texas only corporations may qualify.[13]

Property Tax Exemptions. In general, county and municipal governments derive most of their revenue from taxes on real property and fixed improvements on that property. One popular tax incentive enacted by many states is the allowance of property tax exemptions for innovative, energy-efficient improvements, including solar heating and cooling systems, wind energy conversion systems, and other sources of renewable energy. Without the property tax exemption, improvements to the farm operation might be reassessed and the tax burden increased. Property tax exemptions reduce the effective cost of investing in energy-saving technology.

By far the most common piece of tax incentive legislation has been the exemption of solar equipment from state or local property taxes, in whole or in part. Over twenty states provide a property tax exemption. Some permit counties, cities, or townships to exempt solar devices; others exempt the difference in value between solar and conventional systems; and still others totally exempt solar heating and cooling systems.[14]

Sales Tax and Use Tax Exemptions. A few states have exempted from the state sales tax and use tax the purchase of energy-saving equipment, typically solar energy systems.[15] Of particular interest to agricultural policymakers are the Arizona laws that exempt from the transaction privilege tax and use tax solar energy devices designed to provide heating or cooling, to produce electrical or mechanical power, or to pump irrigation water.[16] The effect of these tax exemptions is, of course, to reduce the cost of investing in energy-conserving devices.

Proponents of tax incentives cite four advantages. First, tax incentives appear to be the most politically acceptable policy tool available, being less coercive than taxes or regulations. Second, it is argued that tax incentives are necessary to make investment in energy conservation economically attractive. Third, it is claimed that tax incentives require less government involvement and are simpler to administer than direct expenditure programs. And fourth, because farmers are familiar with the tax aspect of farming, some proponents claim that tax incentives will foster private decisionmaking.

Policymakers have apparently embraced tax incentives as the favored method of encouraging energy conservation. The popularity of this approach, unfortunately, seems to be based on a widespread lack of appreciation of the practical, equitable, and institutional disadvantages of tax incentives. These disadvantages are formidable and should be considered before any legislation is passed.

As a practical matter, it is difficult to define eligible investments to exclude those that would have been made in the absence of any incentive, yet include those that would not have. Since it is virtually impossible to determine whether an investment was made for the purpose of energy conservation, tax incentives will provide windfalls to some investors, if eligibility is broadly defined, or will have limited usefulness, if eligibility is narrowly defined. The use of tax incentives also encourages investment in machines and equipment, but does nothing to promote labor-intensive techniques, although the latter may be more energy-efficient.

Many critics argue that tax incentives create inequitable social results. In general, the benefits of tax incentives are unevenly distributed. Consider, for example, two farmers interested in making identical purchases of eligible energy-conserving, solar-powered irrigation pumps. Farmer A runs a successful operation and has a large taxable income; farmer B is hard put to show any taxable income. Income tax rates are generally progressive, and a tax deduction is more beneficial to farmer A than to farmer B, although farmer B probably has a more pressing need for the tax benefit. In addition, policymakers must remember that tax incentives are like tax expenditures in terms of fiscal budgeting. At least in the short term, tax breaks result in a revenue decline and a redistribution of the tax burden onto other taxpayers. Moreover, tax incentives are open to criticism as unnecessary giveaways when they are used to encourage farmers to invest in something that will ultimately save them money.

When a tax incentive is under consideration, it is difficult to predict who will use the incentive or what the cost will be. Once enacted, a tax incentive is difficult to control, since it usually does not appear as part of the annual budget review and, unlike spending programs, is not presented to the governor for signature or veto. Typically, the cost of an incentive gets hidden from both policymakers and the public, and an examination of its cost-effectiveness is nearly impossible.

There is probably no tax incentive program that cannot be redesigned as a direct government expenditure—in the form of a loan, grant, or technical assistance—to avoid these disadvantages. Yet it appears likely that energy conservation will continue to be encouraged by the use of tax incentives. Consequently, states should consider placing conditions on whatever tax incentive is chosen to maximize its effectiveness and minimize its cost. Some possible conditions include:

- Placing a dollar ceiling on the amount of the tax allowance to limit revenue losses to the state;

- Making the tax incentive effective immediately and setting a specific termination date, to discourage delay and to limit the ultimate cost of the program;

- Allowing farmers to carry forward any unused tax benefit to future tax years or, alternatively, refunding the excess benefit to extend the tax break to farmers who are temporarily not making a profit;

- Permitting a property tax exemption only until such time as the tax savings equals the initial energy-saving investment.

If states should choose tax incentives as a way to stimulate investment in energy conservation, legislators must still define what types of investments qualify. As a policy matter, tax incentives should probably be offered only to farmers who invest in innovative approaches that hold substantial promise of energy savings but are financially risky. Tax breaks should probably not be allowed for investment in technologically proven low risk investments, despite their capacity for reducing energy consumption. These are legitimate political questions that deserve full discussion as each state decides how to stretch its limited budget to achieve the greatest energy saving for each dollar spent.

Legal Limitations
These incentives may not be feasible in every state, due to legal requirements that may restrict the expenditure of public funds or the allowance of tax breaks for energy conservation. The most important limitation, adopted by almost every state, is the "public purpose" requirement. Simply stated, states and cities may tax and spend public funds only to serve the public, not private, interest.[17] The difficult legal question arises when a public expenditure—for instance, a state energy conservation loan to a farmer for insulating a dairy barn—has both a public and a private purpose. Such a loan reduces energy waste and increases the amount of fuel available to everyone, two benefits that accrue to the general populace. But such a loan also reduces the borrowing farmer's financing cost and fuel bills, two distinctly private benefits.

The most commonly used test for determining whether a public expenditure violates the public purpose requirement is to compare the benefits received by the private party with the benefits received by the state. To pass the public purpose requirement, the primary benefit must accrue to the people of the state rather than to the individual farmer. It appears almost certain that energy conservation incentives can be brought within the scope of acts undertaken for a public purpose. Energy conservation in the production of food and fiber can produce vitally important benefits to all the people of a state or local community, because reduced consumption of non-renewable resources frees fuel for other important uses and reduces air and water pollution.

Since courts tend to give substantial weight to legislative findings, the most important step that legislative drafters can take to assure that an energy conservation program will meet the public purpose requirement is to specifically enumerate its public purpose and the expected benefits to the state. A more complicated alternative is to pass a constitutional amendment permitting the state to expend public funds for projects that produce private benefits.

A second significant legal limitation contained in many state constitutions forbids state and local governments from offering credit to aid any individual, corporation, or association.[18] Some states also prohibit the expenditure of public funds on grants, gifts, or donations to private persons.[19] Thus, credit clauses may restrict a state from offering energy conservation loans, demonstration grants, or free technical assistance to farmers.

Third, many state constitutions limit the amount of debt or liability that a state or locality may incur.[20] In some jurisdictions, these debt limitations may restrict the power of government to subsidize investment in energy conservation with loans or grants. These debt ceilings may be less confining than they appear; one commentator has concluded that "most states now find themselves able to borrow for public improvements of any nature, constitutional restrictions notwithstanding."[21]

Finally, some state constitutions declare that all taxation shall be uniform and prohibit the allowance of any tax exemptions except those enumerated in the constitution.[22] These uniformity clauses restrict the power of the state to permit property and income tax exemptions to farmers who invest in energy conservation. These restrictions can be overcome only by passing a constitutional amendment, such as the one recently adopted in Georgia permitting counties and cities to exempt solar heating and cooling systems from ad valorem taxation.[23]

In sum, state legislators and local officials would do well to determine the limits of their authority before recommending expenditures of public funds in the name of energy conservation.

TAXES

The power to tax can serve not only to raise revenue, but also to influence energy conservation decisions.[24] A large, although unpopular, step could be taken toward reducing waste in agriculture if states taxed the purchase of energy-intensive refined petroleum fuels, natural gas, electricity, fertilizers, and pesticides. Naturally, using taxes to implement energy policy demands careful calibration and fine tuning if the goal of energy conservation is to be met without unwanted side effects. In many cases, a combination of taxes, spending programs, and regulation will pave the most prudent path toward energy conservation.

Taxes are a sure way to alter behavior. For their impact to be felt, however, taxes on energy-intensive inputs would have to be large— even large enough to effectively double the price of energy. Taxes of this magnitude are likely to raise political hackles as well as serious legal questions.

The first of these legal questions is whether state taxes may be applied for regulatory purposes. The U.S. Supreme Court answered it in *Magnano Co.* v. *Hamilton*, in which a tax levied by the state of Washington on butter substitutes was upheld in spite of the obvious intent to suppress the use of margarine.[25] A later case, which sustained a hefty federal tax on marijuana, reiterated Supreme Court approval of regulation through taxation, as the Court declared that "[i]t is beyond question that a tax does not cease to be valid merely because it regulates, discourages, or even definitely deters the activities taxed. . . ."[26] These cases seem to promise clear passage for state taxes designed to discourage energy waste on farms by boosting the cost of energy-intensive fuels and other materials.

A second legal problem with state taxing powers crops up in the commerce clause of the U.S. Constitution, which prohibits states from enacting taxes that discriminate against or unduly burden interstate commerce. The courts have evolved a series of guidelines to determine the legality of a proposed state tax on goods that move in interstate commerce. Such a tax stands only if three conditions are fulfilled:

1. The taxpayer has a certain minimum level of contact with the taxing state;
2. The burden on the taxpayer is commensurate with the benefits received by the taxpayer from the taxing state; and
3. The collective burden placed on the taxpayer by all states imposing taxes does not obstruct interstate commerce.

Within these confines, two types of taxes can be enacted to provide disincentives to energy waste in agriculture: an energy sales tax and an energy users tax.

Energy Sales Tax

Many states presently grant a tax exemption when natural gas, gasoline, electricity, fertilizers, and pesticides are sold to farmers. Now that these commodities are becoming ever more scarce, perhaps such tax exemptions are no longer justified. Rescission of at least a portion of these exemptions would stimulate energy efficiency and discourage waste.

In general, any state sales tax levied on a sale consummated within state borders is valid, even if the goods are produced in another state. Imposing a sales tax on energy-intensive agricultural materials where none now exists and lifting sales tax exemptions that do exist offer certain advantages. First, such legislation will likely result in reduced energy waste, as farmers react to higher prices on agricultural materials by stepping up their efforts to conserve energy. Second, the administration of a sales tax is simple, with the seller of the product collecting the tax at the point of sale. Utilities could easily tack on a sales tax based on the dollar amount of energy consumed, and fertilizer dealers could collect a sales tax along with the price. Third, a sales tax would bring net gains in revenue that can go to fund other energy conservation programs. Unlike spending programs that cause state budgets to balloon, sales taxes on energy-intensive agricultural materials fatten state coffers.

Of course, disadvantages common to all taxes limit the application of the sales tax to agriculture. One clear consequence of such a sales tax would be increased costs of production. Recent years have brought escalating costs for fuels, fertilizer, and agricultural machines: farmers are already hard pressed. Unable to pass these higher costs on to consumers, at least in the short run, farmers would be forced to dip into their profits. Today, many small farms that produce food in an energy-efficient manner are squeezed by increasing production costs and uncertain commodity prices. These farmers can ill afford any loss in profits.

A tax on energy-intensive production inputs also creates the potential for decreased production as farmers cut back their use of taxed agricultural materials. According to one projection, doubling the price of nitrogen fertilizer and irrigation water (two energy-intensive inputs responsible for a sizeable share of current food production) would suppress the use of each by 14 percent and 28 percent, respectively.[27] In addition, taxes that increase energy costs will certainly appear as higher food prices in the long term. A doubling of energy prices would lead to an estimated average commodity price hike of 13 percent.[28] The figure for retail food prices could only be greater, since further links in the food marketing chain—processing, transportation, and retailing—are even more energy-intensive than food production. Finally, a sales tax on agricultural materials may not be the most effective type of tax to induce energy conservation. Since it is levied on sales price rather than on amount of energy consumed, a sales tax is less directly related to energy conservation than is an energy users tax.

Energy Users Tax

States could take a more direct approach by taxing energy users on the amount of energy consumed. Often called a BTU tax, a general energy tax can control demand for all forms of energy, inducing conservation of energy in all phases of food production. Such a tax applies directly to the consumption of primary fuels (e.g., oil, natural gas, coal, and uranium) and indirectly to commodities (e.g., electricity and fertilizers) whose production consumes primary fuels.

Probably the simplest approach to administering a BTU tax would be to collect it at the point of extraction—at the wellhead or the mine shaft. But if only energy-producing states collect the tax, the other states are bound to object. A more equitable alternative would be to tax those who use energy initially. Thus, in the food system, states could assess the diesel fuel that powers tractors, the natural gas that drives irrigation pumps, and the LP gas that dries grain. A state assessment on electric utilities could be based on the amount of energy contained in the coal, oil, or uranium that fires the boilers. Manufacturers of fertilizers and pesticides could be taxed as well on the energy content of the fuels used in their production.

Again, a tax on energy would have to be stiff, perhaps enough to double the price of energy, if demand is to be significantly reduced. Of course, such a potent tax dose would have to be administered with care to minimize undesirable effects. States could couple the BTU tax with antidotes in the form of rebates or exemptions to relieve economic hardship on individual farmers.

Assessing users for the BTUs they consume offers three advantages. First, an energy tax will effectively encourage conservation, since reducing energy waste will cut production costs. Second, states will find the BTU tax relatively easy to administer, since tax collection can occur at highly visible points: fuel distributors, utilities, and manufacturers. A combination of accurate metering and careful monitoring will guarantee that energy users are fairly assessed. Third, revenues from the BTU tax could be used by states to subsidize other programs designed to promote energy conservation.

The disadvantages of an energy users tax echo those of the energy sales tax: increased production costs for farmers and consequent profit losses, reduced crop yields as farmers cut back on fertilizer application and irrigation, and eventual increased food prices as costs are passed along to consumers.

In sum, strong political and practical forces operate against the imposition of energy taxes, but taxes have the appeal of being an equitable method of encouraging energy conservation.

REGULATIONS

State regulation of private farm activity stems from the "police power"—a legal term describing the inherent power of a government to act in its discretion to promote the public health, safety, morals, and general welfare.[29] These four interests—and the concept of the general welfare in particular—define the enormous territory subject to police power regulation. The Supreme Court has described the police power as one of the least limitable of governmental powers and has acknowledged its flexibility to meet the complex needs of a changing society.[30]

Although states are accorded wide latitude in choosing how to promote the general welfare, state regulation intended to foster energy conservation must conform to specific standards in order to receive judicial approval. Energy conservation legislation must have as its goal a legitimate objective; it must bear a reasonable relation to the attainment of that objective; and it must not be arbitrary or unreasonable.[31]

Since state regulation to promote energy conservation must have a proper objective, the first question to be asked is whether the goal of increased energy efficiency in agricultural production justifies governmental intervention. Policymakers should have little reason to fear that energy conservation will fail to pass this test. Just as public concern over the fate of our physical environment brought environmental objectives within the scope of police power regulation, the growing interest in conserving energy should also receive the same acceptance. Energy conservation should certainly join the list of general welfare goals that have become the focus of broadened public concern. Recent federal, state, and local legislation designed to protect Americans from energy shortages leaves little doubt that conserving energy is a legitimate objective under even the most stringent construction of the scope of the police power.

State regulatory legislation intended to enhance energy conservation must be reasonably related to achieving that legitimate objective. As a rule, state legislatures are given considerable leeway in fashioning legislation to remedy problems of public concern. However, energy conservation legislation whose primary purpose is not to save energy but to effect some other goal not authorized by the police power will be struck down.

State regulation cannot be arbitrary or unreasonable. Regulation that has no likelihood of achieving the goal of energy conservation or that creates an unreasonably harsh effect on farmers or their property will not withstand legal challenge. In general, those who seek to overturn state regulation will assert that the regulation is an impermissible taking of property without due process of law, that it is preempted by federal legislation and contravenes the supremacy clause of the Constitution, or that it creates an undue burden on interstate commerce. Before considering each of these potential legal obstacles, it may be useful to describe some of the energy-conserving regulatory strategies that have been suggested for agriculture.

Possible Regulatory Strategies

The following list of regulatory strategies is intended to offer state agricultural policy planners an idea of the range of regulatory options available. The list is intended to be thought provoking; it does not purport to be exclusive, nor does inclusion on the list mean that each strategy is equally effective or desirable. The legal difficulties of enactment and enforcement and the practical difficulty of assessing the impact on agricultural production are considered in subsequent sections of this chapter.

1. *Regulate the end use of fuels.*   States could prohibit the use of a particular fuel for certain energy-inefficient farm activities.

2. *Require farmers to install energy-conserving materials or to adopt energy-saving methods.*   For example, states might, when appropriate, require farmers to adopt limited tillage methods or to retrofit livestock buildings with insulation.

3. *Impose quotas or restrictions on the availability of energy and other natural resources.*   If a critical energy supply shortage occurs again, as it did in some states during the winter of 1976–1977, states could limit the availability of fuels, electricity, and energy-intensive irrigation water.

4. *Impose mandatory energy conservation targets.*   States could adopt a more flexible method of restricting resources by setting mandatory energy conservation targets, requiring farmers to cut energy consumption by a certain percentage. To increase the viability of this method, states might impose economic penalties on those who failed to achieve the targeted goal.

5. *Increase the prices of fuel and other natural resources through state regulatory action.* States, through their public utility commissions, are empowered to regulate the prices of electricity, intrastate natural gas, and water. Raising the price of these commodities could reduce energy consumption on farms, as well as in other sectors of the economy.

6. *Require farmers to have farm machinery and buildings inspected annually for energy efficiency.* Annual energy audits can help farmers pinpoint sources of energy inefficiency in farm operations, providing information that is useful in deciding whether to repair or replace energy-inefficient equipment. Inspection of farm vehicles could be coupled with a registration fee schedule that charged more for energy-inefficient vehicles and less for energy-efficient ones.

7. *Require manufacturers of farm vehicles or equipment to incorporate specified energy-conserving items.* Radial ply tires, electronic ignition, and fuel consumption meters on tractors and other farm vehicles can help conserve energy. Legislation could mandate that any farm vehicles or equipment sold in the state must be equipped with these (or other) energy-conserving features.

8. *Require manufacturers of farm vehicles or equipment to label these items for energy efficiency.* A state could require manufacturers to label for fuel efficiency their tractors, crop dryers, and other farm machinery. The use of standardized testing and labeling can provide information that will be used for comparison shopping by prospective buyers. Farm equipment could also be labeled for recommended operating methods that can enhance fuel efficiency.

Legal Issues

There is no doubt that states possess the power to allocate and conserve scarce material resources upon and beneath their land.[32] Nor is there a dispute that states can regulate how energy is produced and used. As the Supreme Court, speaking through Justice Brandeis, wrote, ". . . the [l]egislature may, for the purpose of conserving natural resources, regulate their production and use."[33]

The possible regulatory approaches suggested in the preceding list will, most likely, spark considerable controversy among agricultural policymakers and farmers. States that choose to regulate the availability and price of fuel, to prohibit or require the use of particular farming methods, to mandate annual inspections for energy efficiency, or to prescribe standards for farm equipment manufacturers should expect their actions to be challenged. The purpose of this section is to analyze the legal challenges that state regulations are likely to encounter.

Taking. Whenever a state restricts farming activities or the use of property without providing compensation, legal challenges to the regulation as a taking of private property are possible. The federal Constitution and state constitutions prohibit states from depriving a person of property without just compensation.[34]

The issue, of course, is whether state regulation equals a taking. It is fair to say that this question is one of the most difficult that exists in the law. The test most commonly applied is a balancing test that weighs the benefit derived by the public against the burden placed upon the person whose activity or property is regulated. Is it reasonable for a state to prohibit a farmer's uses of fuel for particular purposes, to establish restrictions on fuel availability, or to set up mandatory energy conservation targets? Can a state reasonably require farmers to adopt new farming techniques or to modify (retrofit) their existing stock of energy-consuming equipment? Can states require farmers to have their farm vehicles inspected annually to determine energy efficiency?

To balance the competing interests of the state and the individual, the Supreme Court has articulated the following standard:

> To evaluate [the regulation's] reasonableness, we therefore need to know such things as the nature of the menace against which it will protect, the availability and effectiveness of other less drastic protective steps, and the loss which appellants will suffer from the imposition of the ordinance.[35]

A court must first consider the nature of the interest that the legislation seeks to protect. Regulations to protect the community health and safety are accorded great weight compared to the burden placed on the individual. Other interests are more difficult to categorize. For example, regulations to protect the local economy or to preserve the environment may be ranked differently by different courts. It is difficult to predict how regulations geared to conserve energy will be judged. Of course, the answer to the question is not fundamentally a legal one, but rather a judicial expression of the tenor of the times about the limits that ought to be placed on the government.

In considering whether regulation constitutes an impermissible taking without just compensation, the regulation need not cause the property to be physically taken. At some point, regulation will diminish the value of property to such a degree that courts will deem the regulation unreasonable. The Supreme Court agreed that "[t]he general rule . . . is that while property may be regulated to a certain extent, if regulation goes too far it will be recognized as a taking."[36] No simple formula exists for calculating the allowable diminution in value. Although it is an important factor, diminished value alone will not be decisive except where the property becomes unuseable and all value is lost.

Prohibiting the use of certain fuels for particular farming operations could create problems, depending on the availability of alternative energy sources and the economic losses sustained by individual farmers. If farmers were required to install new equipment or to adopt new methods to comply with fuel use controls, courts might find that such regulation was unreasonable and an unconstitutional taking. On the other hand, a regulation that retroactively required private owners to spend a large sum to comply with newly created building standards has been upheld as valid.[37]

Because each case presents different facts, a prediction of how each court will weigh competing interests is impossible. The gravity of the energy problem seems to suggest that state policymakers will be given leeway in fashioning a cohesive regulatory policy for agriculture. Legislators, nevertheless, would be wise to regulate in such a way as to temper any harsh effects on individuals.

Pre-emption. The supremacy clause of the Constitution makes federal legislation the supreme law of the land, and state regulation that is incompatible with federal law must acquiesce.[38] There are no easy tests for determining when courts will decide that a state law has been pre-empted by federal legislation.

In general, states are free to act in areas where Congress has been silent. When both the state and federal governments have acted to regulate the same activity, however, a challenge to state regulation on the basis of federal pre-emption is possible. Coincidental regulation by both federal and state governments is permitted, and states are free to fill in the gaps left by federal regulation, unless the state law directly conflicts with federal law, the operation of the state law interferes with or impairs the federal regulatory scheme, or Congress has intended federal regulation to be exclusive.

When a direct conflict exists between a state law and federal law that makes it impossible to comply with both, clearly the state law must yield.[39] The Supreme Court has declared that "[a] holding of federal exclusion of state law is inescapable and requires no inquiry into congressional design where compliance with both federal and state regulation is a physical impossibility for one engaged in inter-state commerce."[40]

Instances of direct conflict are rare. Most pre-emption challenges arise when both the state and federal governments have taken similar actions to fulfill similar goals. Unless a court is convinced that the operation of the state law seriously hinders a clear federal policy, both laws will be harmonized, and coincidental state regulation will be permitted. State laws may be struck down, however, where they will impair a significant federal policy. For example, a state statute that imposed stricter controls on nuclear power than did the federal regulations was invalidated, because the more stringent state law impaired the federal purpose of encouraging the development of nuclear power.[41]

Congress can pre-empt state regulation simply by expressing its intent to do so. This intent may be found in specific statutory language, or it may be implied in the nature of the federal legislation. When congressional intent is silent, the outcome of a pre-emption challenge to state regulation is uncertain. Courts have found implied congressional intent in areas where coincidental regulation might have produced inconsistent results, where the subject matter of federal legislation demanded nationwide uniformity, and where the federal interest was deemed dominant.

State agricultural policy planners should face few pre-emption problems in enacting carefully drafted energy conservation legislation. Requiring manufacturers to label farm vehicles and equipment for fuel consumption or for recommended operation methods to increase fuel efficiency would withstand a pre-emption challenge, since these requirements supplement those of the federal Energy Policy and Conservation Act of 1975.[42] States could also enact controls on the price and availability of most fuels without running afoul of federal laws that pre-empt state action. States must exercise restraint, however, if they attempt to reduce demand for natural gas by raising its price or limiting its availability. The boundary designating what states may and may not regulate is confused, and any regulation of natural gas not clearly confined to intrastate commerce is likely to invite claims of state intrusion into an area pre-empted by federal legislation.[43]

Commerce Clause. The U.S. Constitution has given Congress the exclusive right to regulate trade and commerce among the states.[44] Based on the belief that free-flowing internal trade is in the national interest, the commerce clause provides a significant limitation to the power of the states to legislate energy-conserving regulatory measures. The Constitution makes it clear that when Congress has acted, all contrary or frustrating state laws must yield. Still, the troubling question remains: In the absence of congressional action which state regulations are valid?

Beginning with the case of *Cooley* v. *Board of Wardens*,[45] decided over one hundred years ago, the Supreme Court has attempted to cope with this problem. In *Cooley*, a Pennsylvania law required all ships to hire local pilots to guide them into and out of Philadelphia harbor. Cooley, who was fired for violating this statute, argued that the law constituted a burden on interstate commerce. The Court disagreed, drawing the distinction between "national" concerns that require a uniform system a regulation and "local" problems that are best provided for by the states. Determining whether a problem is local or national in scope is not always simple. In practice, courts have attempted to balance the national interest in free interstate commerce against a state's interest in protecting the health and safety of its citizens. Thus, state regulation of goods before they actually move into the flow of interstate commerce has been upheld, while state statutes that adversely affected the movement of interstate trains have been invalidated.[46]

Even if a problem is purely local, state regulations are not permitted to act in such a way to discriminate against goods that flow in interstate commerce. Thus, state regulations cannot bear more heavily on out of state businesses than on local businesses through the imposition of different tax rates or different licensing provisions. What appears to be the current test for balancing the burden on interstate commerce against the state's legitimate interest in the health, safety, and general welfare of its citizens was recently stated by the Supreme Court in *Pike* v. *Bruce Church, Inc.*[47] In *Pike*, the Court invalidated an Arizona requirement that cantaloupes had to be crated in order to be shipped out of the state, declaring:

> [w]here the statute regulates even-handedly to effectuate a legitimate local public interest, and its effects on interstate commerce are only incidental, it will be upheld unless the burden imposed on such commerce is clearly excessive in relation to the putative local benefits.[48]

Will state energy conservation regulations withstand the challenge that they unduly burden or discriminate against interstate commerce? Whether a particular state energy conservation measure—requiring tractor manufacturers to install specific energy-saving devices, for example—will be upheld may well depend on persuading a court that the measure is both effective and reasonable. States can take steps to maximize the likelihood that their energy conservation regulations will withstand such a challenge.

First, legislation should emphasize the connection between the regulation and the health and safety of the people in the state. State legislators should, to the extent possible, document the need for energy conservation and demonstrate how the particular piece of legislation can enhance the state's economy. Second, legislation should describe the state's particular need for regulation and outline how the chosen approach creates a minimum of interference with interstate commerce. Third, state policymakers should strive to create regulations that take the least onerous approach. And finally, legislation should not favor local products, businesses, or services, nor should legislation be enacted that invites retaliation from other states.

Enabling Authority. All powers not delegated to the federal government nor prohibited to the states are reserved to the states by the tenth amendment to the Constitution. However, a political subdivision of the state, such as a county or city, can enact energy conservation ordinances only if the state grants it the authority to do so through enabling legislation.

One way to grant a local government the power to act in areas of municipal concern is the legal device of "home rule," which grants municipalities broad powers to act for the general welfare of their residents. Local governments granted home rule may act within the same framework as states to promote the general welfare. Localities without home rule authority must rely on the more narrowly based conferral of power granted by specific state enabling provisions.

Local governments have a role to play in helping conserve energy in food production. As more and more people take up home gardening and livestock raising in our cities and suburbs, local governments may want to become involved in encouraging or regulating urban food production. States should ensure that local governments are granted the necessary power to minimize any legal challenges to local action.

Impact of Regulation
The major advantage of state and local regulation of energy use is that it can be put into effect with little delay and can lead to an immediate reduction of energy consumption. The effectiveness of natural resources regulation is attested to by the reduced use of natural gas after the imposition of emergency state regulations in the winter

of 1976—1977 and by the cutback in water consumption following the use of water conservation regulations in Marin County, California, and in the suburbs of Washington, D.C.

Nevertheless, many of the suggested regulatory strategies create pitfalls that may far outweigh their potential for saving energy. Restricting the availability of energy or greatly increasing its price would probably cause a reduction in agricultural output as farmers cut back on irrigation and fertilizer use.[49] Restricting energy availability or increasing energy prices is also likely to increase food costs. A 10 percent restriction in energy availability could lead to a 55 percent commodity price increase; the result of doubled energy prices, though not as intense, could result in a 13 percent commodity price hike.[50] The potential for reduced output, increased food prices, and pared farmer profits should serve to put any astute policy planner on notice that the politics of regulation are fraught with danger.

Prohibiting the use of fuel for particular purposes or requiring the adoption of energy-saving methods may also be difficult to enforce. The regulatory strategies that seem least offensive—requiring annual inspections of farm machinery, compelling manufacturers to install energy-saving devices on farm equipment, and mandating manufacturers to affix energy efficiency labels on farm equipment—offer fewer problems, but the payoff in energy savings is not easily measured. Furthermore, regulation, more than any other type of state program, is likely to run afoul of legal limitations.

In sum, regulatory strategies to conserve energy have a part to play in a well-balanced state program, but care should be exercised to avoid unwanted repercussions. Strategies that permit a gradual transition will permit farmers to upgrade their farms and to amortize the costs of current equipment over a period of time. The result will be a more humane and less painful conversion to new agricultural methods that squeeze the last drop of value out of a gallon of fuel and that wring the last unit of work out of a kilowatt-hour of electricity.

INFORMATION TRANSFER PROGRAMS

There is a shortage of information about energy conservation, a shortage to which energy waste in agriculture can be at least partly attributed. Where it occurs, communication between agricultural researchers and farmers is often spotty. Farmer X may be unaware that his favorite tractor is energy-inefficient. Farmer Y may wish to keep abreast of energy-saving technology improvements, but does not have the time. Farmer Z may seek an expert's help in deciding whether her plan to put in a new, energy-efficient irrigation pump is economically sound. Increasing the flow of reliable information about how to conserve energy on farms is an important first move that can only lead to more intelligent use of energy in the future. States can gather and spread information through a variety of research, demonstration, and educational programs.

Collection

Agricultural research is the first step. Despite budgetary problems, agricultural research is underway in a curious composite of federal agencies, state institutions, and private organizations spanning the nation.

The hub of the research system is the U.S. Department of Agriculture (USDA), which plans, funds, and staffs a multitude of research projects involving energy, food, and fiber. Most agricultural research programs are cooperative efforts in which both state and federal governments take part. Within each state, one or more State Agricultural Experiment Stations (SAESs) conduct research under the general guidance and funding of the federal Cooperative State Research Service, an agency within the USDA. Other agencies within USDA that conduct and co-sponsor research projects concerning energy consumption and conservation in agriculture include the Agricultural Research Service and the Economic Research Service. Additional research is conducted by the land-grant colleges and by private research organizations.

Efforts have been made to keep researchers in one place current on diverse projects underway elsewhere and to coordinate overall research activities. USDA's computerized Current Research Information Service provides instant access to information about the research activities of the six USDA agencies, fifty-three SAESs, and twenty-five additional cooperating state institutions. The Agricultural Research Policy Advisory Committee (ARPAC) comprises top level administrators of the USDA, representatives of SAESs at the land-grant colleges, and other members interested in the direction of agricultural research. The most recent attempt to coordinate and disseminate research, specifically solar energy research projects, will be undertaken pursuant to the Food and Agriculture Act of 1977.[51]

Today, researchers are studying how renewable energy sources (including solar, wind, and biomass) can be effectively used to reduce fossil fuel consumption in agriculture. Although much of the funding comes from the federal government, most of the actual research is carried out by nonfederal researchers.[52] Some states, notably Montana and New Mexico, have budgeted state revenues to fund energy research. Montana devotes 2.5 percent of its coal severance tax revenue to research, development, and demonstration of alternative energy sources.[53] This subsidy amounted to approximately $500,000 in 1976 and will probably increase when the percentage allocated to research and development rises by law to 5 percent in 1980. New Mexico has set aside funding of at least $500,000 for solar energy research and development, information and education, and equipment testing and standardizing.[54] Other states, especially those where agriculture constitutes a major component of the economy, could make similar commitments of state funds to agricultural research programs geared toward energy conservation.

Private agricultural research organizations provide another source of technological creativity. The federal government recognizes the value of private consultants and research groups, and several federally funded research projects are underway. Lockheed Aircraft, for example, is studying the use of solar energy for reducing energy consumption in greenhouse operations in a program funded by ERDA (now a part of the Department of Energy) and administered by USDA.[55] Development Planning and Research Associates is analyzing the use of wind energy on farms for USDA.[56] And the federal Community Services Administration currently supports the Small Farm Energy Project's three-year study designed to investigate the feasibility of implementing energy conservation techniques and proven renewable energy innovations on small, low income farms in Nebraska.[57]

Just as the federal government contracts with private consultants and researchers, states could offer research grants on a competitive basis.[58] Such funding of agricultural research in the private sector can benefit states in two ways. States gain a degree of flexibility in approaching agricultural problems. Depending on the nature of the project and on the expertise required, work can be done in-house or contracted out. In addition, private research groups often can offer fresh perspectives. Outside the established community of government and university researchers, where old biases and prejudices may reign, new researchers are more apt to be inventive in approaching agricultural problems from a different angle.

Dissemination
Researchers in any field of technology are familiar with the flood of articles that bursts forth annually, making it difficult to keep abreast of current research activity. Besides funding and staffing research programs, states should energetically publicize the results of research projects.

Cooperative Extension Service. The fruits of agricultural research have long been disseminated via the outreach program of the popular Cooperative Extension Service (CES), operated jointly by the USDA and by states at the land-grant universities.[59] Since its inception in 1914, the CES has steadily spread technical farming information over all of America. Although national in scope, the CES is local in flavor. Its network of county agents, state specialists, and state university researchers provides specialized information tuned to the needs of the individual farmers served. Contact between CES county agents and farmers has always been direct, underscoring the local focus of the entire outreach program.

The programs offered by the CES can be broadly categorized as active, reactive, and educational.[60] Active programs address general problems brought up by state or federal staff, reactive programs deal with specific problems identified by individual farmers, and educational programs transmit general information and stimulate farmers' awareness of new ways to increase profits.

An active program might come to life when, for example, CES staff recognizes that a lack of information about pest populations is preventing farmers from making sufficiently informed decisions about pesticide application. By establishing a statewide system to monitor pests, gathering research reports and other data on the subject, and distributing all findings to the farmers, the CES would try to fill in the information gap. One example of this is found in Michigan, where a statewide apple pesticide advisory service developed at the state university synthesizes all available information to decide when pesticide application is most efficacious, then alerts apple growers via a telephone hotline.[61] The result has been an impressive savings in energy-intensive pesticides.

Farmers spark a reactive response from CES personnel by asking a specific question. For instance, a farmer might ask for help in deciding what sort of irrigation system would most benefit corn grown on sandy soil. Or, an extension service expert might be invited to a dairy farm operation to give advice on how insulation, solar energy equipment, and waste heat exchangers could cut both energy consumption and production costs.

General education programs encompass the many activities through which county agents advise farmers how to conserve energy and reduce production costs. Among these activities are classes, seminars, workshops, and field days.

What can be done to improve the collection and distribution of information on conservation in agriculture? States can marshal past experience and present knowledge to expand and supplement CES energy information transfer programs in at least six ways.

First, states should evaluate the adequacy of CES funding. Currently, 45 percent of funding comes from the federal government, 37 percent from states, 16 percent from counties, and 2 percent from other sources.[62] A stepped up budget would allow the hiring of additional experts to direct the expansion of such state energy conservation efforts as energy audits of farm operations, more outreach programs focusing on energy efficiency in farming, and additional demonstration projects.

Second, states should intensify the emphasis on direct contact between CES agricultural energy experts and farmers. Face to face is a most effective way to transmit information. Direct meetings can fit a number of formats, including classes, seminars, workshops, and individual consultations. Farmers can get an illustrative, often startling slant on their energy consumption from individual energy audits. Such audits pinpoint where energy is used on farms, where it is wasted, and what can be done to eliminate waste.

Third, states can make sure that CES personnel are equipped to answer energy-related requests of individual farmers. Certainly any farmer seeking advice on how to improve farm procedures deserves a prompt and thorough response. States must make sure that extension service personnel are adequate in training and in number. Only then can energy conservation be effected at all levels.

Fourth, farmers can best be convinced to use energy-saving alternatives by observing their effectiveness under realistic conditions. State demonstration projects could work several ways. A state-developed and funded demonstration program could be similar to the one set up by the New York State Energy Research and Development Authority to study the feasibility of using wind energy on New York dairy farms.[63] Naturally, state-funded demonstration projects can be expensive. Perhaps one solution to the fiscal problem is for states to co-sponsor demonstration projects, such as the solar-powered irrigation system recently dedicated near Albuquerque, New Mexico, a project jointly funded by the state of New Mexico, ERDA, and the Four Corners Regional Commission.[64] Private funding for demonstration projects may also be possible. Battelle Memorial Institute is developing a demonstration fifty-horsepower solar-powered irrigation pump in Gila Bend, Arizona, in a project funded by the Northwestern Mutual Life Insurance Company.[65] Some other suggestions for state demonstration programs include showing off new energy-saving methods on state experimental farms and having private farmers host demonstration projects, with the understanding that they will inherit the new equipment once the project is completed.[66] Similarly, states could pay farmers already practicing energy-efficient methods to allow other farmers in the area to observe them.

Fifth, a state outreach program should answer financial as well as technological questions raised by farmers. Typical information transfer programs stress the technological side of the problem. But farmers need to see the economic side of the picture, too. Fears that financial returns on investments in energy-efficient equipment may be insufficient to cover changeover costs can seriously block the progress of conservation projects. To help quell such fears, states should provide full information concerning the economic costs and benefits of various investments. Pertinent data would include projected cost savings, payback periods, and tax advantages.

Sixth, information provided by extension services must be timely and readily available. Delays defeat the purpose of an information service. Telephone hotlines could get the right information to farmers at the right time—when they need it.

Energy Extension Service. One idea that has garnered recent federal attention is the creation of a national energy extension service (EES), modeled after the popular Cooperative Extension Service. Congress has passed and the president has signed into law the National Extension Service Act of 1977, under which grants totaling approximately $11 million have been awarded to implement two-year pilot programs in ten states.[67] As envisioned, the EES will use its funds to encourage the adoption of techniques and technologies that save energy or use renewable energy sources. To achieve these objectives, the focus of the program will be the provision of direct assistance and information to small energy users, in a form that will be both personalized and convenient. Some of the information to be covered includes the availability, technical details, and energy- and cost-saving potential of innovative energy techniques and technologies. The goal of the Energy Extension Service is to create a credible network of information transfer that will speed the adoption of energy-efficient techniques.

On its own, Georgia has enacted a law establishing a state energy extension service program to provide information and technical assistance relating to energy conservation measures, energy-efficient technologies, and available alternate energy technologies.[68] As part of its duties, the Georgia EES must provide technical assistance, advisory services, public education and training workshops, and a feedback mechanism to maintain awareness of energy research and development needs at the local level.

Fuel Efficiency Labeling. States should require farm machinery manufacturers to include the fuel-efficiency rating on labels. Without this information, comparison shopping is difficult. Currently, automobile manufacturers must label each passenger car for its fuel consumption based on standardized tests, and a similar requirement could be legislated for tractors, combines, irrigation pumps, crop dryers, and other farm equipment.

Fortunately, not all machinery manufacturers would have to start from scratch, since standardized tests for tractor fuel consumption and efficiency have already been developed. For many years, the University of Nebraska tractor tests have measured fuel consumption under standardized conditions.[69] The results of these tests are compiled and published annually.

The only conceivable legal argument against fuel-efficiency labeling—that federal action has pre-empted the field—is not likely to be a problem. The federal Energy Policy and Conservation Act of 1975 sets standards for energy-efficiency labeling only for passenger automobiles and for other consumer products, which do not include farm machinery.[70] This argument thus should pose no barrier to the establishment of individual state energy-efficiency labeling requirements.

CONCLUSION

This chapter has set out the four categories of strategies available to policymakers concerned with conserving energy in agriculture: financial incentives, taxes, regulations, and information transfer programs. Financial incentives and information transfer programs are the least coercive of these options, since they force no one to obey against his or her will. They are probably the most effective methods in terms of political acceptability, ease of enforcement, and ability to withstand legal challenge. Energy taxes and regulations offer alternative methods to encourage or require energy conservation, but care must be taken to avoid strategies that are counterproductive. A well-designed plan for agricultural energy conservation is likely to be a mix of strategies taken from some or all of the four categories. Given the wide range of options, it is up to thoughtful agricultural policymakers to select that combination most appropriate to the economic, social, and political conditions of the state.

NOTES TO CHAPTER 2

1. Much of the following analysis is drawn from another book in this series. *See* Norman L. Dean, *Energy Efficiency in Industry* (Cambridge, Massachusetts: Ballinger Publishing Co., forthcoming).

2. Mont. Rev. Codes Ann. § 84−7402(3) (Supp. 1977).

3. Md. S.B. 742 and H.B. 1598 (1976) (Not enacted).

4. 42 U.S.C.A. § 6326(4) (West Supp. 1977).

5. Mont. Rev. Codes Ann. §§ 89−3604; 84−7411 to 7412 (Supp. 1977).

6. Mont. Rev. Codes Ann. § 89−3603 (Supp. 1977). Although the Montana law does not specifically include the term "energy," it is likely that a state loan would be available to a farmer who wants to develop renewable energy resources, such as solar irrigation pumps or methane digesters. Because of its relevance to farmers and ranchers, the text of this section is set out below.

89−3603. Renewable resource development loans. (1) The board of natural resources and conservation is authorized upon proper application and upon recommendation of the department of natural resources and conservation to make loans from the renewable resource development account established by this act to farmers and ranchers of the state of Montana who, without regard to their form of business organization:

(a) are citizens of the United States and are citizens and residents of the state of Montana;

(b) have sufficient farming or ranching training and experience which, in the opinion of the department, is sufficient to assure the likelihood of the success of the proposed operations; and

(c) are or will become owner-operators of farms or ranches.

(2) The department shall solicit and consider in its evaluation of proposed projects the views of interested and affected departments, boards, agencies and other subdivisions of state and federal government and of other interested and affected persons.

(3) The board may make the renewable resource development loans provided for by this section for any worthwhile project for the conservation, management, utilization, development, or preservation of the land, water, fish, wildlife, recreational, and other renewable resources in the state; and for the refinancing of existing indebtedness incurred in the expansion or rehabilitation of projects for those purposes.

(4) The board shall make no renewable resource development loan which exceeds the lesser of one hundred thousand dollars ($100,000), or eighty per cent (80%) of the fair market value of the security given therefor. In determining the fair market value for the security given for any loan, the department shall consider appraisals made by qualified appraisers and such other factors it considers important.

(5) The period for repayment of loans pursuant to this act may not exceed thirty (30) years.

(6) The board shall from time to time establish by rule the interest rate at which loans may be made under this act, provided that in no case may the rate be greater than one (1) percentage point greater than the prevailing interest rate on the renewable resource development bonds provided for in this act.

(7) The state shall have a lien upon a project constructed with money from the renewable resource development account for the amount of the loan, together with the interest thereon. This lien may attach to all pro-

ject facilities, equipment, easements, real property and property of any kind of nature owned by the debtor, including all water rights. The board shall file either a financing statement or a real estate mortgage covering the loan, its amount, terms and a description of the project with the county recorder of each county in which the project or any part thereof is located. The county recorder shall record the lien in a book kept for the recording of liens and it shall be indexed as other liens are required by law to be indexed. The lien shall be valid until paid in full or otherwise discharged. The lien shall be foreclosed in accordance with applicable state law governing foreclosure of mortgages and liens.

(8) The board may adopt rules as required to govern the terms and conditions for making loans, security instruments, and agreements pursuant to this act.

(9) No member, officer, attorney, or other employee of the board or the department shall, directly or indirectly, be the beneficiary of or receive any fee, commission, gift, or other consideration for or in connection with any transaction or business under this act other than such salary, fee, or other compensation as he may receive as such member, officer, attorney, or employee. Any person violating any provision of this section shall, upon conviction thereof be punished by a fine of not more than two thousand dollars ($2,000) or imprisonment for not more than two (2) years or both.

(10) The department shall administer the loans made by the board pursuant to this act, and may accept and utilize voluntary and uncompensated services, and, with the consent of the agency concerned, utilize the officers, employees, equipment, and information of any agency of the federal government, or of any agency of Montana government, or of any political subdivision within Montana.

7. Mich. H.B. 4465 (1975) (Not enacted).

8. 7 U.S.C. §§ 1501–1520 (1970 & Supp. V 1976).

9. Ch. 114, 1977 N.M. Laws (to be codified in N.M. Stat. Ann. § 72–15A–11.4).

10. Kan. Stat. Ann. § 79–32,167 (Supp. 1976).

11. *See, e.g.*, Minn. H.F. 1905 (1976) (Not enacted); Minn. H.F. 2141 (1976) (Not enacted); Minn. H.F. 1718 (1975) (Not enacted). For example, one proposed bill would have provided

[a] credit of ten percent of the net cost of equipment and devices approved by the Minnesota energy agency that are purchased, installed, and operated within Minnesota exclusively to produce fuel or electric power from organic residues from livestock feeding facilities may be deducted from the tax due . . . in the taxable year in which such equipment is purchased or installed. . . .

Minn. H.F. 1905 (1976) (Not enacted).

12. Ariz. Rev. Stat. Ann. § 43–123.37 (Supp. 1977). In addition, Ariz. Rev. Stat. Ann. § 43–128.03 (Supp. 1977) provides taxpayers with the option of taking tax credits for the installation of solar energy devices, in lieu of depreciation or amortization deductions.

13. Kan. Stat. Ann. § 79–32,168 (Supp. 1976); Tex. Tax-Gen. Ann. art. 12.01(6) (Vernon Supp. 1976).

14. These states now include Arizona, Colorado, Connecticut, Georgia, Hawaii, Illinois, Indiana, Maine, Maryland, Massachusetts, Michigan, Montana, Nevada, New Hampshire, New York, North Carolina, North Dakota, Oregon, Rhode Island, South Dakota, Virginia, Vermont, and Washington.

15. *Sales tax exemptions*: Ariz. Rev. Stat. Ann. § 42–1312.01(A)(9) (Supp. 1977); Ga. Code Ann. § 92–3403a(C)(2)(z.1) (Supp. 1977); Me. Rev. Stat. Ann. tit. 36, § 1760(37) (Supp. 1977), *added by* ch. 542, § 4, 1977 Me. Acts; Mich. Comp. Laws Ann. § 205.54h (Supp. 1977); Tex. Tax-Gen. Ann. art. 20.04(CC) (Vernon Supp. 1976).

Use tax exemptions: Ariz. Rev. Stat. Ann. § 42–1409(B)(9) (Supp. 1977); Ga. Code Ann. § 92–3403a(C)(2)(z.1) (Supp. 1977); Mich. Comp. Laws Ann. § 205.94e (Supp. 1977).

16. Ariz. Rev. Stat. Ann. §§ 42–1312.01(A)(9), 42–1409(B)(9) (Supp. 1977).

17. *See, e.g.*, Tex. Const. art. VIII, § 3.

18. *See, e.g.*, Wis. Const. art. VIII, § 3; Idaho Const. art. VIII, §§ 2, 4.

19. *See, e.g.*, Ala. Const. art. IV, § 94.

20. *See, e.g.*, Wis. Const. art. VIII, § 7; Idaho Const. art. VIII, § 3.

21. A. James Heins, *Constitutional Restrictions Against State Debt* (Madison: University of Wisconsin Press, 1963), p. 83. *See also* G. Theodore Mitau, *State and Local Government: Politics and Processes* (New York: Charles Scribner's Sons, 1966), p. 606.

22. *See, e.g.*, Tex. Const. art. VIII, § 1; Wis. Const. art. VIII, § 1.

23. Ga. Code Ann. § 2–4604 (1977).

24. For a more complete discussion of how taxes and financial disincentives can be used to reduce energy consumption, *see* Joe W. Russell, Jr. (with the assistance of E. Grant Garrison), *Waste Not, Want Not: Energy Conservation Through Economic Disincentives* (Cambridge, Massachusetts: Ballinger Publishing Co., forthcoming).

25. 292 U.S. 40 (1934).

26. United States v. Sanchez, 340 U.S. 42, 44 (1950).

27. Dan Dvoskin and Earl O. Heady, *U.S. Agricultural Production Under Limited Energy Supplies, High Energy Prices, and Expanding Agricultural Exports* (Ames: Center for Agricultural and Rural Development, Iowa State University, 1976), p. 14.

28. *Ibid.*, p. 78.

29. License Cases, 46 U.S. (5 How.) 504, 582 (1847). Some of the following discussion of regulations and legal issues is based on another book in this series. *See* Corbin Crews Harwood, *Using Land to Save Energy* (Cambridge, Massachusetts: Ballinger Publishing Co., 1977), pp. 214–19.

30. Village of Euclid v. Ambler Realty Co., 272 U.S. 365, 386 (1926); Queenside Hills Realty Co. v. Saxl, 328 U.S. 80, 83 (1946).

31. Goldblatt v. Town of Hempstead, 369 U.S. 590 (1962).

32. *See, e.g.*, Walls v. Midland Carbon Co., 254 U.S. 300 (1920); Bandini Petroleum Co. v. Superior Court, 284 U.S. 8 (1931).

33. Henderson Co. v. Thompson, 300 U.S. 258, 267 (1937).

34. "[N]or shall any State . . . deprive any person of . . . property, without due process of law." U.S. Const. amend. XIV, § 1. State constitutions contain similarly worded provisions prohibiting uncompensated takings of property.

35. Goldblatt v. Town of Hempstead, 369 U.S. 590, 595 (1962).

36. Pennsylvania Coal Co. v. Mahon, 260 U.S. 393, 415 (1922).

37. Queenside Hills Realty Co. v. Saxl, 328 U.S. 80 (1946).

38. "This Constitution, and the Laws of the United States which shall be made in Pursuance thereof; and all Treaties made, or which shall be made, under the Authority of the United States, shall be the supreme Law of the Land; . . . any thing in the Constitution or Laws of any State to the Contrary notwithstanding." U.S. Const. art. VI, cl. 2.

39. Gibbons v. Ogden, 22 U.S. (9 Wheat.) 1 (1824).

40. Florida Lime & Avocado Growers, Inc. v. Paul, 373 U.S. 132, 142–43 (1963). The Court found no such conflict in this case and upheld the state law.

41. Northern States Power Co. v. Minnesota, 447 F. 2d 1143 (8th Cir. 1971), *aff'd per curiam*, 405 U.S. 1035 (1972).

42. 15 U.S.C. § 2006; 42 U.S.C. §§ 6294, 6296 (Supp. V 1976).

43. Federal Power Comm'n v. Corporation Comm'n of Oklahoma, 362 F. Supp. 522 (W.D. Okla. 1973), *aff'd*, 415 U.S. 961 (1974); Northern Natural Gas Co. v. State Corp. Comm'n of Kansas, 372 U.S. 84 (1963).

44. U.S. Const. art. I, § 8, cl. 3.

45. 53 U.S. (12 How.) 299 (1851).

46. *See, e.g.*, Parker v. Brown, 317 U.S. 341 (1943) (regulation of raisins packaged prior to interstate commerce upheld); Southern Pacific Co. v. Arizona, 325 U.S. 761 (1945) (state regulation limiting train length invalidated).

47. 397 U.S. 137 (1970).

48. *Id.* at 142.

49. *See* Dvoskin and Heady, *supra* note 27 at 139.

50. *Ibid.*, p. 92.

51. To collect the necessary information, the Food and Agriculture Act of 1977 states that

> The Secretary shall, through the Cooperative State Research Service and other agencies within the Department of Agriculture which the Secretary considers appropriate, in consultation with the Energy Research and Development Administration, other appropriate United States Government agencies, the National Academy of Sciences, and private and nonprofit institutions involved in solar energy research projects, by June 1, 1978, and by June 1 in each year thereafter, make a compilation of solar energy research projects related to agriculture which are being carried out during such year by Federal, State, private, and nonprofit institutions and, where available, the results of such projects. Such compilations may include, but are not limited to, projects dealing with heating and cooling methods for farm structures and dwellings (such as greenhouses, curing barns, and livestock shelters), storage of power, operation of farm equipment (including irrigation pumps, crop dryers and curers, and electric vehicles), and the development of new technologies to be used on farms which are powered by other than fossil fuels or derivatives thereof.

7 U.S.C.A. § 3251 (West Supp. February 1978).

52. The USDA Agricultural Research Service has, for example, funded university researchers to study the use of solar energy for dairy operations in Arizona, *Solar Energy Intelligence Report* 2, no. 15 (July 19, 1976): 120; for poultry housing in Maryland, *ibid.* 1, no. 13 (September 1, 1975): 102; and for greenhouses in Indiana, *ibid.* 2, no. 17 (August 16, 1976): 136.

In Kansas and Nebraska, an eighteen-month, $300,000 FEA-funded project is underway to develop a model for voluntary statewide energy conservation in farm production. The goal of the program, which includes over 200 cooperating farmers from more than twenty-five counties in each state, is to reduce by 15 to 20 percent the amount of energy used in production agriculture without adversely affecting crop yields. Although federally funded, the project is being carried out by state personnel. The agricultural engineering departments of the University of Nebraska at Lincoln and Kansas State University at Manhattan are performing the field work; the Nebraska Energy Office and the Kansas State Energy Office are acting as liaison agencies.

Further information and progress reports are available from:

Lyle E. Goltz
Assistant Director for Conservation/Allocation
State of Kansas Energy Office
503 Kansas
Topeka, KS 66603
(913) 296−2496

George J. Dworak
State Allocations Officer
Nebraska Department of Revenue
Box 94818
Lincoln, NE 68509
(402) 471−2867

53. Mont. Rev. Codes Ann. § 84−1319(2)(b) (Supp. 1977).
54. Ch. 347, 1977 N.M. Laws.
55. *Solar Energy Intelligence Report* 1, no. 13 (September 1, 1975): 103.
56. *Ibid.* 2, no. 22 (October 25, 1976): 175.
57. *Small Farm Energy Project Newsletter* no. 4 (Hartington, Nebraska: Small Farm Energy Project, Center for Rural Affairs, February 1977), p. 1.
58. The development of state competitive grant programs could take the form of the recently enacted law under which the secretary of agriculture will award competitive grants for research and development relating to

(1) uses of solar energy with respect to farm buildings, farm homes, and farm machinery (including, but not limited to, equipment used to dry or cure farm crops or forest products, or to provide irrigation); and
(2) uses of biomass derived from solar energy, including farm and forest products, byproducts, and residues, as substitutes for nonrenewable fuels and petrochemicals.

7 U.S.C.A. § 3241 (West Supp. February 1978).

59. 7 U.S.C. §§ 341–349 (1970, Supp. V 1976, and West Supp. February 1978).

60. This analysis borrows from the work of H. LeRoy Marlow, director of the Pennsylvania Technical Assistance Program. *See Oversight Hearings on Intergovernmental Dissemination of Federal Research and Development Results Before the Subcomm. on Domestic and International Scientific Planning and Analysis of the House Comm. on Science and Technology*, 94th Cong., 1st Sess. 164 (1975).

61. *See* "Michigan Farmers Use Phone for Advice on Spraying," *New York Times*, August 1, 1976, p. 22.

62. Federal Council for Science and Technology, *Directory of Federal Technology Transfer* (Washington, D.C.: U.S. Government Printing Office, 1975), p. 4.

63. *Solar Energy Intelligence Report* 3, no. 22 (July 11, 1977): 148.

64. *Ibid.* 2, no. 24 (November 22, 1976): 140; *ibid.* 3, no. 23 (July 18, 1977): 152.

65. *Ibid.* 2, no. 24 (November 22, 1976): 140; *ibid.* 3, no. 12 (May 2, 1977): 88.

66. Precisely this approach has been taken in the Food and Agriculture Act of 1977, which provides for the establishment of model farms and demonstration projects to research the feasibility of solar energy on farms. The act describes the establishment, operation, and activities of these model farms and demonstration projects in the following way:

§ 3261. Model farms—Establishment and operation.

(a) In order to promote the establishment and operation of solar energy demonstration farms within each State, the Secretary shall distribute funds to carry out the activities described in subsections (b) and (c) of this section and section 3262 of this title to one or more of the following in each State: the State department of agriculture, the State cooperative extension service, the State agricultural experiment station, forestry schools eligible to receive funds under the Act of October 10, 1962, or colleges and universities eligible to receive funds under the Act of August 30, 1890, including Tuskegee Institute (hereinafter in this part referred to as "eligible institutions"), in accordance with such rules and regulations as the Secretary may prescribe.

Activities and responsibilities of fund recipients
(b) The recipient or recipients in such State shall—
(1) establish at least one large model farm which—
(A) demonstrates all the solar energy projects determined by by the Secretary, in consultation with the recipient or recipients, to be useful and beneficial to the State;
(B) is located in the State on land owned or operated by that State and, if practicable, on the State agricultural experiment station farm land; and
(C) includes other farming practices, such as raising livestock and crops, in order to provide a model of a farm which applies new and improved methods of agriculture through the use of solar

energy as a means of heating, cooling, drying, or curing crops, and providing other farm needs;

(2) sell the products of the model farm established under paragraph (1) of this subsection and pay to the Secretary that portion of the proceeds received through each such sale as bears the same proportion to the total proceeds as the grants under this section bear to the total cost of operating the farm. The Secretary shall deposit such funds into a fund which shall be available without fiscal year limitation for use in carrying out the provisions of this part;

(3) provide tours of the model farm to farmers and other interested groups and individuals and, upon request, provide such farmers, groups, and individuals with information concerning the operation of such model farm and the demonstrations, if any, established by it under section 3262 of this title;

(4) determine the costs of energy, the income, and the total cost of the model farm; and

(5) annually compile a report concerning energy usage, income costs, operating difficulties, and farmer interest with respect to the model farm and the demonstrations, if any, established by it under section 3262 of this title, and submit the report to the Secretary along with any recommendations concerning project changes and specific needs of such farm or demonstrations.

Extension of results

(c) The results obtained from each model farm established under subsection (b) of this section which prove to be economically practical shall be extended to other farms in each State through the State cooperative extension service as part of its ongoing energy management and conservation education programs.

§ 3262. Demonstration projects—Establishment

(a) During each calendar year after the first two calendar years for which eligible institutions in a State receive grants pursuant to section 3261 of this title the recipient or recipients of such grants in each State, in consultation with the Secretary, shall establish not less than ten demonstrations of solar energy projects which they shall select from among the projects demonstrated on the model farm established in the State pursuant to section 3261 of this title. Such demonstrations shall be carried out on farms which are already operating in the State.

Activities and responsibilities of fund recipients

(b) The recipient or recipients in each State shall enter into written agreements with persons who own farms and who are willing to carry out solar energy project demonstrations under this section. Such agreements shall include the following provisions concerning solar energy projects which the owners agree to demonstrate on such farms:

(1) The owner shall carry out the projects on the farm for such period as the Secretary determines to be necessary to fairly demonstrate them.

(2) Tools, equipment, seeds, seedlings, fertilizer, equipment, and other agricultural materials and technology which are necessary to carry out the projects and which, on the date of such agreement, are not commonly being used on farms in such State, shall be provided by the recipient or recipients.

(3) During the demonstration period, the recipient or recipients, with the assistance of the Extension Service of the Department of Agriculture, shall provide the owner with technical assistance concerning such projects.

(4) During the demonstration period and for such other periods as the recipient or recipients deem necessary, the owner shall—

(A) keep a monthly record for the farm of changes, if any, in energy usage and costs, the amount of agricultural commodities produced, the costs of producing such amount, and the income derived from producing such amount, and of such other data concerning the projects as the recipient or recipients may require; and

(B) transmit to the recipient or recipients such monthly records, along with a report containing his or her findings, conclusions, and recommendations concerning the projects.

(5) During the demonstration period, the owner shall give tours of the farm to farmers and other interested groups and individuals and provide them with a summary of the costs of carrying out such projects.

(6) All right, title, and interest to any agricultural commodity produced on the farm as a result of the projects shall be in the owner.

(7) At the end of the demonstration period, the owner shall have all right, title, and interest to any materials and technology provided under paragraph (2) of this subsection.

(8) Such other provisions as the Secretary may, by rule, require.

7 U.S.C.A. §§ 3261, 3262 (West Supp. February 1978).

67. 42 U.S.C.A. §§ 7001–7011 (West Supp. September 1977). The ten states chosen were Alabama, Connecticut, Michigan, New Mexico, Pennsylvania, Tennessee, Texas, Washington, Wisconsin, and Wyoming. U.S. Energy Research and Development Administration, *Information from ERDA* no. 77–134 (August 5, 1977).

68. Res. 29, 1977 Ga. Laws.

69. *See* University of Nebraska, Department of Agricultural Engineering, *Nebraska Tractor Test Data* (Lincoln: University of Nebraska, College of Agriculture, 1974).

70. 15 U.S.C. § 2006; 42 U.S.C. §§ 6294, 6296 (Supp. V 1976).

Chapter 3

Reducing Direct Uses
of Energy on Farms

Since the Second World War, the nature of agricultural production in the United States has changed dramatically, as farmers have adopted new technologies that rely more heavily on fossil fuel energy than on human and animal labor. In the United States, the agricultural revolution has had profound effects on farm population, acreage cultivated, and crop yields. Comparing farm sector demographics of 1950 and 1973 will help illustrate some of these changes:

- The number of farms was halved, from 5.6 million to 2.8 million.
- The number of farmworkers decreased from 9.9 million to under 4.4 million.
- Harvested acreage decreased from 345 million acres to 322 million acres while, at the same time, the acreage used to produce export commodities increased from 50 million acres to 100 million acres.[1]

The notable social changes that have accompanied the agricultural revolution are partially a result of vastly increased farm productivity. During this time, the crop yield per acre increased more than 60 percent and output per man-hour increased almost fourfold, as the productivity of both the farmers and the land improved.[2] Thus, fewer farmworkers tilling less land produced more food for more people. As former Secretary of Agriculture Clifford Hardin wrote:

> Using a modern feeding system for broilers, one man can take care of 60,000 to 75,000 chickens. One man in a modern feedlot can now take

care of 5,000 head of cattle. One man with a mechanized system can operate a dairy enterprise of 50 to 60 milk cows.

Agriculture, in short, does an amazingly efficient job of producing food
for an ever larger number of people.[3]

Few will dispute the basic thrust of that statement, but many observers of modern agriculture note that the farmer who raises these
chickens, cattle, and dairy cows does so only with the aid of many
other workers who may never set foot on the farm. To state that one
farmworker can feed over fifty persons, as is frequently claimed,[4] is
misleading hyperbole, because it ignores the farmer's dependence on
vital inputs from many support industries. Indeed, many of the 6
million farmworkers who were displaced from the farm over the last
three decades now produce the inputs that are the very basis for the
modern farm, and it is estimated that there are two farm support
workers for every farmworker.[5]

The key to the American farmer's impressive level of efficiency,
as measured in output per man-hour, is the shift from muscle power
to fossil fuels. Production agriculture today depends on farm machinery, refined petroleum products, fertilizers, pesticides, and electricity. The technological revolution on the farm has resulted in
profound changes, leading one observer to state that "agriculture
today has shifted in a very real sense from soil to oil."[6]

This chapter will focus on how energy is directly used on farms in
the production of food and fiber. This portion of energy use in agricultural production includes, by definition, the direct application of
petroleum products and electricity to various tasks on the farm.
Thus, the chapter examines the use of energy to operate farm machinery, irrigate crops, dry crops, heat greenhouses, and protect
orchards from frosts.

Several studies have attempted to estimate the level of direct energy consumption on farms, and although the exact figures vary, the
conclusions are quite similar. A recent study of energy use in the
food system, prepared for the Federal Energy Administration, estimates that the energy directly expended to produce food for domestic consumption amounts to almost 690 trillion BTUs, or about 1
percent of all energy consumed in the United States.[7] This figure
does not include the large use of energy to produce nonfood crops,
such as cotton and tobacco, nor does it consider energy used to produce all the crops that are exported. If these are included, the direct
consumption of energy on farms is about twice as great: approximately 1,375 trillion BTUs, or 2 percent of total energy consump-

tion in the United States.[8] Other researchers attempting to measure the energy consumed on farms reach similar results.[9]

Agriculture is both a consumer and a producer of energy. As a consumer of energy, agriculture uses more petroleum products than any other single industry in the nation.[10] Still, total energy consumption on farms—both direct and indirect—amounts to only 3 to 4 percent of the energy used annually in the nation. When one considers that farmers use this energy to produce the crops that feed Americans and many people of the world, such an expenditure can be considered neither extravagant nor overwhelming. It is nevertheless substantial, and farmers, like everyone else, share the responsibility to conserve energy. The goal, of course, is to do so in a way that is technologically, economically, politically, and nutritionally sound. And that is the goal of the remainder of this chapter: to provide options for change that will meet these criteria.

FARM MACHINERY OPERATIONS

One of the most important causes of increased agricultural efficiency, in terms of output per man-hour, has been the mechanization of American farms. Tractors and other farm machinery, however, guzzle large quantities of gasoline and diesel fuel to propel themselves across the land planting, spraying, cultivating, fertilizing, and harvesting. Today, farm machinery—including tractors, combines, trucks, and other power implements—is the single largest user of fuel and energy in agricultural production, accounting for almost three-fifths of all energy consumed.[11]

Although the number of acres of cropland used for crops remained relatively stable between 1940 and 1975, tractors proliferated. The number of tractors on farms rose from just under 1.6 million tractors in 1940 to a peak of 4.8 million in 1965, before beginning a gradual decline to a 1975 level of 4.2 million.[12] The abundant supply of inexpensive petroleum products made mechanization of the farm inevitable on economic grounds alone. Horses eat every day, and their food requires land for its production; this land could be used to produce cash crops that more than offset the costs of the machinery and fuel. Farmers also found that using less human labor and more machinery increased profits. As a result, farmers increased their capital expenditures for farm machinery and equipment from $625 million in 1940 to $8.8 billion in 1975.[13] Indeed, the value of farm machinery stock in 1975 rose to $46.5 billion, more than double that of 1965.[14] At the same time, the human labor required for farming decreased from over 20 billion man-hours in 1940 to only 5.5 billion man-hours in 1974.[15]

As a result of this mechanization, the amount of fuel consumed by farm machinery for farm operations was almost 3.7 billion gallons of gasoline and 2.5 billion gallons of diesel fuel in 1973, about 800 trillion BTUs.[16] Moreover, if one includes the fuel consumed by automobiles for farm-related operations, an additional 360 million gallons of gasoline, representing 45 trillion BTUs, are burned.[17] These estimates may be low. Economist Michael Perelman, working from different data, estimates that tractors alone consumed over 8 billion gallons of fuel in 1972, or about 1 quadrillion BTUs.[18] Perelman states that this amount exceeds the energy value of the foods we consume, a turnabout from the days when agriculture was a net energy-producing sector of the economy.

The Council for Agricultural Science and Technology (CAST) calculates that improved operation, maintenance, and design of farm equipment could, in the aggregate, reduce fuel use by 21 percent.[19] Some of the suggested improvements to increase energy efficiency include converting to farm machinery that is powered by diesel fuel rather than gasoline, improving machinery operation and maintenance, and redesigning machinery.[20] Moreover, improved tillage practices such as "no-till" and minimum tillage could conserve additional fuel energy.[21]

Convert to Diesel-Powered Farm Machinery

The conversion from gasoline- to diesel-fueled machinery offers the single greatest potential for fuel savings in farm machinery operations, an estimated 643 million gallons of gasoline a year.[22] This represents an energy savings of over 80 trillion BTUs, or 10 percent of all fuel consumed by farm machines. From an energy standpoint, diesel-fueled machinery is preferable because it is more efficient than a gasoline-powered engine. A gallon of diesel fuel yields about 10 percent more energy than a gallon of gasoline, and due to higher engine compression, the same amount of work can be done with about 27 percent less fuel if diesel machinery is used. From the farmer's perspective, diesel-fueled engines are advantageous because they are more economical to operate. Diesel fuel generally costs less per gallon, and diesel engines require less maintenance.

The economics of owning and operating diesel machinery make investment in it a wise decision. Recognizing this, many farmers are shifting to diesel-powered tractors and combines on their own initiative. In 1973, for example, 86 percent of all new tractors sold were diesel, compared with 81 percent in 1972. In the same year, nearly

50 percent of new self-propelled combines were diesel, compared with 35 percent in 1972 and only 23 percent in 1971.[23] This trend is considered likely to continue in the future.

If the marketplace is operating to make farmers convert to diesel-powered tractors and combines, one might ask why states should intervene at all. What are the problems that states must consider in determining whether to pursue policies that encourage farmers to purchase diesel machinery or that discourage them from purchasing gasoline machinery?

First, policymakers should understand that tractors and other heavy machinery have long useful lives and that the equipment bought today is going to be in operation for many years. While the stock of diesel tractors and the proportion of diesel tractors to gasoline tractors continues to increase, it is projected that over 1.5 million gasoline tractors will still be operating in 1980.[24] Thus, strategies that accelerate conversion to diesel machinery today will have energy-saving consequences for many years. Second, some farmers continue to purchase gasoline machinery instead of diesel machinery with the same horsepower due to the higher initial cost of diesel farm machinery. In 1973, the list prices for a seventy-horsepower tractor averaged about \$8,240 for diesel and \$7,410 for gasoline.[25] The price paid for the average tractor purchased today is much greater, as a consequence of generally higher prices[26] and a shift to tractors with greater horsepower.[27] For example, between 1963 and 1975 the average size of tractors sold increased from fifty-five to ninety-seven horsepower.[28]

States that want to accelerate the trend toward more diesel-powered farm machinery can pursue several strategies:

1. Provide Loans for Investment in Energy-Saving Diesel Farm Machinery. One way for states to lower the barrier of high initial cost is to offer financial support, making capital investment loans available at favorable interest rates to farmers who purchase new diesel equipment or who retrofit old gasoline-powered machinery with new diesel engines. State programs could take the form of direct state loans to farmers, state loan guarantees to private lenders, or interest subsidies.

One way to allocate limited loan funds is to provide a loan for an amount equal to the cost differential between comparably powered diesel and gasoline machinery. By limiting the loan to the cost difference between the two types of engines, a state will be able to stretch its budget to help more farmers. States should, however, consider whether it is a good idea to provide financial resources to help farmers purchase energy-efficient, cost-saving machinery whose purchase presents no economic risk. Probably the key argument against making such loans is the fact that farmers are presently converting to diesel machinery on their own, without government intervention.

2. Provide Tax Incentives for Investment in Diesel Farm Equipment. To encourage investment in energy-conserving farm equipment, a state could offer tax incentives to farmers who buy new diesel machinery or who retrofit their old gasoline machinery with diesel engines. A state could permit an investment tax credit, allowing farmers who purchase diesel equipment to credit a percentage of the purchase price against any income tax they owe. A state could allow farmers to depreciate diesel equipment at an accelerated rate, thereby reducing the effective cost of purchasing diesel machinery. Or a state could exempt the purchase of diesel machinery from any applicable sales tax, adding an incentive to farmers who are weighing the costs and benefits of purchasing diesel machinery.

As with loan programs, states must carefully consider the wisdom of offering tax breaks to encourage farmers to invest in equipment that is cost-effective. Diesel equipment has proven itself to be economically advantageous for large farm tasks, and it is questionable whether states should offer additional incentives.

3. Impose Taxes That Favor the Use of More Energy-Efficient Diesel Machinery. Farmers can be encouraged to convert to diesel machinery or discouraged from operating gasoline vehicles by the imposition of taxes that favor the use of diesel equipment. For example, a state could assess a special tax on gasoline that is not charged on diesel fuel. The effect of such a surtax on gasoline would be to increase its price compared with diesel fuel, making diesel fuel and diesel machinery more appealing economically. A second alternative is for a state to assess a higher registration and licensing fee for gasoline machinery than for diesel machinery. Some states currently impose higher registration fees for heavier weight passenger cars than for lighter weight cars. A state could similarly assess a higher registration fee for gasoline-powered farm vehicles than for diesel equipment, to account for the energy impact of the less energy-efficient gasoline vehicles.

Tax policies that favor diesel over gasoline machinery will, of course, increase the costs of purchasing and operating gasoline machinery compared with diesel machinery. Although the imposition, collection, and enforcement of such taxes should present no unusual administrative difficulties, states must recognize that the imposition of any new taxes will probably arouse substantial opposition.

Improved Operation and Maintenance

CAST estimates that improved operation and maintenance of farm machinery can save almost 346 million gallons of gasoline and 278 million gallons of diesel fuel.[29] This represents a savings of over 82 trillion BTUs, which is about 10 percent of all energy expended in operating farm machinery.

Vehicle efficiency can be improved, for example, by tuning farm machinery at regular intervals. On the average, this would result in a 10 to 15 percent improvement in engine performance. If all operators followed good maintenance procedures, the savings could amount to roughly 5 percent of current fuel consumption.[30]

There are numerous operating techniques that can increase fuel efficiency. Matching the equipment to the job to be done can save energy. This technique is not always possible, however, especially for the smaller family farms that own one large tractor to handle all the operations for which a tractor is needed. Many farmers wonder what they can do to improve the energy efficiency of their tractors when they are used for light jobs such as disking, harrowing, planting, or spraying. Several studies have shown that large tractors pulling light loads operate more economically if farmers shift to a higher gear and throttle down the engine. The engine then runs at a slower speed, and fuel consumption is reduced. CAST reports that farmers utilizing this technique register fuel savings of almost 20 percent.[31] Additional fuel savings are possible if farmers avoid unnecessary idling of their machinery, properly adjust their field equipment, follow optimum traffic patterns through their fields, and select a proper tire tread for the soil type.

Finally, properly ballasting or weighting of tractors for the conditions on a particular farm is important to performance and fuel economy. If a tractor has too much ballast, the tires sink into the soil, which causes soil compaction, increased rolling resistance, and extra fuel consumption. If a tractor has too little ballast, it loses traction, which causes excessive tire wear and tears up the soil. CAST estimates that proper ballasting alone could result in annual savings of 9.6 million gallons of gasoline and 11.7 million gallons of diesel fuel, or almost 3 trillion BTUs.[32]

Several energy-conserving changes in farm machinery operations, along with estimates of the possible percentage of energy savings, are suggested in Table 3–1, which is reported in a recent study sponsored

Table 3-1. **Using Farm Machinery to Save Energy**

• Use the right equipment for the job	5-20%
• Operate tractors near capacity	10-15%
• Use smaller tractors for lighter loads	10-12%
• Move large equipment by truck rather than driving it several miles	8-10%
• Gear up and throttle back on partial loads	5-10%
• Reduce tire slippage by adding wheel weights	6- 8%
• Keep electrical equipment cleaned and lubricated properly	5- 8%
• Use preventive maintenance	4- 6%
• Shut the tractor off rather than let it idle unnecessarily . . .	4- 6%
• Keep all engines tuned up	2- 4%
• Set controls properly	2- 4%
• Couple equipment to reduce trips over the field	2- 4%
• Check equipment manuals for tips on maintenance	2- 3%
• Keep blades sharp	1- 3%
• Double check machinery adjustments	1- 2%
• Clean or replace air filters	1- 2%

Estimated percentage savings based on total expenditure of energy use. Actual savings will vary depending on specific applications, but should fall in these ranges. These savings are not necessarily additive.

Source: G.B. Taylor, "Agricultural Energy Use," in *Efficient Electricity Use*, ed. Craig B. Smith, prepared for the Electric Power Research Institute (New York: Pergamon Press, 1976), p. 399.

by the Electric Power Research Institute. Additional suggestions might include:

1. Provide Education Programs and Advisory Services to Improve Operation and Maintenance of Farm Vehicles. States can provide information to help farmers conserve energy through the outreach programs of the Cooperative Extension Service. For example, state extension specialists and county agents can conduct classes, workshops, and demonstrations that show how proper vehicle maintenance and operation can save energy and money. Giving farmers the most recent information on energy-efficient vehicle operation and maintenance will benefit everyone. For farmers, learning the most current techniques gives them an opportunity to lower production costs and thereby increase profits. For society, increasing the efficiency of fuel consumption on farms will free fuel for other uses.

To help maintain the highest levels of fuel efficiency for farm machinery, states could offer the services of trained experts to inspect and adjust farm equipment for peak operating efficiency. States could use either specially trained state employees or private experts to perform these services. Of course, whether states use their own personnel or contract with private mechanics, there will be some cost. States could, and probably should, charge farmers a reasonable fee for this service. But a reasonable charge need not deter prospective clients if the charge is cost-effective.

In addition to teaching farmers through extension service programs, states can gear education programs to young people, instilling in them an energy conservation ethic. For example, the 4−H program can be used to teach farm youth how improved operation and maintenance of farm vehicles can lead to reduced energy consumption, lower production costs, and increased profits. Another potential source of information is the public school system. Children in rural areas can learn how fuel and energy can be saved in farm machinery operations, through a school curriculum that includes a discussion of energy conservation goals and methods.

2. Require Farm Machinery Manufacturers to Label Vehicles for Fuel Efficiency. As part of a general program to educate farmers about energy consumption and energy efficiency, legislators could require that any farm vehicle sold in the state bear a label telling its fuel efficiency. Although state standard setting for passenger vehicle fuel efficiency labeling is pre-empted by the federal Energy Policy and Conservation Act,[33] no such federal law prohibits states from acting with respect to farm vehicles.

For over fifty years, Nebraska has conducted a comprehensive program of tractor testing. The purpose of the Nebraska Tractor Test is to determine the performance of a tractor, its available horsepower, and its fuel consumption. The results of these tests are disseminated in booklets sold at nominal cost.[34] The results of a standardized test similar to the Nebraska Tractor Test could form the basis of a fuel efficiency labeling program to help farmers make purchasing decisions based on better information.

Fuel efficiency labeling seems to play an important role in consumer decisions regarding passenger automobile purchases, and manufacturers apparently accept the requirement with little, if any, complaint. Indeed, automobile manufacturers now use fuel efficiency ratings in their advertising campaigns to induce a prospective purchaser to buy their particular automobiles. Although farm vehicle manufacturers might complain initially about the labeling requirements, it seems likely that the fuel efficiency of their machinery could become a selling point in persuading would-be purchasers.

Design Changes

Changes in the design of farm equipment could save roughly 2.5 million gallons of gasoline and 23.1 million gallons of diesel fuel, or over 3 trillion BTUs annually.[35] If manufacturers of farm machinery concentrated on energy efficiency, savings of about 5 percent of all energy expended on machine operations could be achieved by 1980.[36]

Farm machines burn energy to power the entire vehicle and all its systems: engine combustion, engine cooling, transmission, air conditioning, and hydraulic controls. Not all of the fuel burned, however, results in useful work, since many of these systems were designed for keeping initial cost low rather than for energy efficiency. Because farm machines have long useful lives, redesigning farm machinery for energy efficiency can return significant fuel savings over time.

To increase vehicle efficiency, a state could require new farm machinery sold in the state to meet minimum standards of fuel efficiency. The federal government has not pre-empted states from acting to set fuel efficiency requirements for farm machinery, and states are free to enact legislation that will govern fuel efficiency of tractors and farm machinery. Although the effect of such a strategy may not be immediate, the long-term impact could be quite substantial, since tractors and other farm machinery have long lives. States could also consider requiring manufacturers of farm machinery to equip new vehicles with such energy-saving items as radial ply tires, electronic ignition systems, improved engine design, and computers to monitor fuel consumption.

Such a strategy may face stiff opposition from farm equipment manufacturers and farmers. Any design changes will entail costs of development and retooling that will be translated into increased purchase prices. Thus, the political opposition to changes of this kind are likely to be similar to the reaction of the automobile manufacturers and many motorists to the requirements for various safety and air pollution control items on new automobiles.

Reduced Tillage

Soil is tilled to improve its physical condition, to eliminate competing vegetation, and to incorporate crop residues. For many hundreds of years, farmers have followed tillage practices that often include plowing in the fall and in the spring, and several diskings and draggings before the crop is planted. In addition, farmers frequently cultivate their fields during the growing season to control weeds. Recent studies show that many crops can be produced economically with little or no tillage, and many farmers are learning that they can save money and time if they abandon the traditional moldboard plow.

Techniques that minimize or eliminate tillage are sometimes referred to as conservation tillage. These techniques include:

- No-till, in which farmers do not plow their fields, disturbing only a two inch strip of soil when they plant the seeds;
- Strip tillage, where farmers limit their tillage to a strip that is one-third or less the distance between rows;
- Till-plant, a system in which farmers scrape the old crop residues and plant the new crop in one operation; and
- Chisel planting, a method that plows only a thin strip of land, leaving residues from the previous crop on the surface to protect the seeds that are planted in the same operation.

When the heavy plowing is minimized or eliminated, fuel and energy are conserved. In addition, reduced tillage also frequently produces other benefits. It decreases the labor necessary to produce a crop, saves time, controls soil erosion, increases soil moisture content, reduces soil compaction, and often results in better yields during dry years. As a result, 35.8 million acres, almost 10 percent of all cropland used for crops, were planted using minimum tillage techniques in 1975.[37] Nearly 6 million acres of this total were prepared using no-till, a scheme that is receiving increasing acceptance by farmers.[38] The popularity of no-till is growing, and the U.S. Department of Agriculture predicts that more than half of all American cropland will be farmed without plowing in thirty years.[39]

The potential energy savings from reducing tillage are enormous. Various studies have reached different results as to the amount of fuel consumed to produce different crops on different soils, but the general conclusion is that reducing the tillage reduces energy consumption. For example, CAST reports a study in which corn grown with the no-till method required only about 3.2 gallons of fuel per acre from planting to harvest, roughly 50 percent less than the conventional tillage methods.[40] This conclusion agrees with a study by the U.S. Department of Agriculture's Economic Research Service, which found that the average preharvest fuel consumption per acre of corn is about 8 gallons for conventional tillage, 6 gallons for reduced tillage systems that use a plow, 4 gallons for reduced tillage systems that do not use a plow, and 2.8 gallons for no-till.[41] Indeed, USDA economist Earle Gavett favorably points to the results of a study of corn tillage systems in Missouri in which conventional practices required 5.5 gallons per acre, reduced tillage 1.4 gallons, and no-till only 0.3 gallons.[42]

The total energy-saving potential of improved tillage practices is not easy to quantify, since tillage requirements vary according to soil, weather, and a variety of other factors. Despite this difficulty, estimates of potential energy savings have been made in two very different states, Texas and Indiana. The Texas study determined that the elimination of unnecessary machine operations and the use of minimum tillage practices could achieve a saving of 10 trillion BTUs, 20 percent of all fuel used for farm machinery operations in that state.[43] The Indiana study of corn production found that conversion to reduced tillage methods could result in an annual saving of almost 7 million gallons of diesel fuel, or about 1 trillion BTUs.[44]

In addition to the direct savings of energy, other benefits can be reaped when tillage is reduced. One of the most important benefits is the savings of time and labor, a consideration that may be more influential than the potential energy savings. For farmers, time is a most valuable asset, and anything that can be done to reduce the time needed to perform a particular operation pays handsome bonuses. Because reduced tillage practices save time during the important spring planting season, farmers have the opportunity to expand their planted acreage. If a farmer's operation is limited by the amount of time and labor needed to till and plant the season's crops, reduced tillage practices can provide the time to plant 10 to 20 percent more acreage.[45]

Minimum tillage, especially no-till, enables the practice of "double cropping"—that is, harvesting a second crop during one year. Abandoning the plow or cutting back on field operations can provide a healthy boost in net income.

One of the major benefits that results from reduced tillage practices is soil conservation. Leaving residues from the old crop impedes the flow of rainfall and reduces soil run-off significantly. One study of the effect of minimum tillage practices on soybean production concludes that soil erosion on no-till cropland is about half as much as it is on conventionally tilled acreage.[46] Using reduced tillage methods will allow farmers to grow crops on sloping fields, opening millions of acres to productive farming that previously had to be kept fallow or in pasture.

In addition, reduced tillage helps retain soil moisture, which may account for the larger yields reported on no-till cropland during dry years. Farmers who irrigate may find that reduced tillage practices are water conserving, with the result that less energy is used to pump irrigation water.

Despite these obvious advantages, farmers and policymakers must recognize some of the disadvantages that go along with reduced tillage methods. First, eliminating or reducing tillage operations may result in a greater profusion of weeds, insects, and rodents. Conventional plowing helps keeps insects and rodents in check; these pests can create greater problems on unplowed fields because they thrive in the undisturbed soil and mulch. Farmers who rely less on the plow may find that they must use greater amounts of energy-intensive chemical herbicides, insecticides, and rodenticides than they did when they plowed the soil. The energy required to produce chemical pesticides is substantial and to some degree offsets the fuel savings that are gained by reducing tractor tillage operations. Nevertheless, most reduced tillage systems use less energy overall than do conventional tillage practices.

Second, minimum tillage techniques are not applicable to all soil types, especially poorly drained or highly organic or claylike soils.

Third, farmers who adopt minimum tillage must become more careful managers of their land. Under conventional systems, farmers make repeated operations, and this redundant system allows mistakes to be corrected. Minimum tillage, on the other hand, reduces the number of operations, leaving less margin for error.

Fourth, the dollar costs of converting to minimum tillage can be expensive, since the new equipment generally costs more than conventional tillage equipment.[47]

Fifth, a frequent criticism of no-till is that it results in short-term crop yield reductions of 5 to 15 percent compared with conventionally tilled fields.[48] Current research, however, indicates that this potential risk of crop loss can be greatly minimized. Indeed, yields on no-till cropland are equal to those of conventionally tilled land in years of sufficient moisture and better than those of conventionally tilled acreage in dry years.[49]

Finally, there is the risk of danger to the environment when increasing amounts of toxic herbicides are introduced into the soil. The effect of increased usage of these chemicals is not fully known, and some agricultural experts feel that this is playing with fire. One must ask whether reduced energy consumption is worth the risk of damage to the environment.

The potential for energy savings is limited to some extent by the reduced tillage practices that many farmers already use. The farmers who grow 75 percent of the corn and a similar share of the soybeans already treat their crops with some herbicides to reduce the number of field operations.[50] In addition, many wheat farmers now practice a form of minimum tillage to retain soil moisture and limit erosion.

Nevertheless, states can follow several courses of action to encourage energy conservation through reduced tillage:

1. Provide Information About the Advantages of Reduced Tillage. Although energy, time, and money can be saved by converting to minimum tillage, many farmers have not yet changed over. One problem faced by an individual farmer is a lack of information about whether reduced tillage will be beneficial. States could help overcome this lack of information by publicizing the potential benefits of minimum tillage practices in general and by analyzing, on request, the farming operation of individual farmers.

For example, one oil company that manufactures herbicides and other agricultural chemicals now offers free advice to any farmer who telephones for information about no-till farming. The Ortho division of Standard Oil of California maintains a toll-free telephone number with experts stationed to answer questions about no-till farming. In addition, Ortho has representatives who will visit a farmer to discuss in detail the use of no-till on his land. Of course, the company is motivated by a desire to make profits and not out of philanthropic concern. States, however, could follow this pattern by offering objective advice from experts within the state department of agriculture or the extension service. The payoff will not be sales or profits, but rather, energy conservation. The cost is the time and money required to send these experts to advise farmers throughout the state. Since each state has different soils and crops, the decision will turn on whether the potential for energy savings outweighs the monetary expense of providing this service.

2. Provide Crop Insurance to Farmers Who Convert to Reduced Tillage. When farmers convert from conventional tillage practices to a system of reduced tillage, they may experience temporary reductions of yield and short-term problems from weeds and other pests. Although the advantages of reduced tillage practices are impressive, farmers may balk at embracing a new tillage technique if there is a risk that profits will suffer. States can minimize the risk of crop yield reduction and profit loss by providing farmers with crop insurance that will guarantee their income as they convert to minimum tillage.

Because farmers are being asked to help society by conserving energy, states in return can help farmers minimize their risk of economic loss. States can, for example, subsidize the cost of crop protection insurance premiums, either totally or partially.

IRRIGATION

Irrigation—the diversion of water onto lands under cultivation—was one of agriculture's earliest technological innovations. Irrigation has been a primary force behind the development of agriculture in many parts of the world in the past and today continues to play a significant role in food production. Water is vital to agriculture, and some observers conclude that in the next twenty-five years the lack of pure, fresh water is likely to become the principal constraint on efforts to expand world food production.[51]

Farmers irrigate their crops for a number of reasons. First, rain falls unpredictably from one year to the next. Irrigation frees farmers somewhat from their dependence on the weather. Second, irrigated land typically yields larger, more stable crops than does unirrigated land. Farmers using irrigation are better able to plan and can expect higher net returns than can the dryland farmer. Third, expanding production by such intensive cropping practices as irrigation is generally cheaper than cultivating new land. Expanded production usually means increased profits, and farmers have striven to augment their irrigated acreage. Fourth, irrigation can bring new lands into production. For thousands of years, farmers have cultivated land that, without irrigation, would have been useless or only marginally productive. The lush desert valleys of the American Southwest and the verdant fields of California's San Joaquin Valley bear dramatic witness to the powers of irrigation. Finally, irrigation makes possible year-round cultivation. In California, many farmers are thus able to grow several crops of lettuce and other vegetables each year to satisfy consumer demand.

These advantages have convinced many farmers to irrigate, and today 54 million acres of land, about 15 percent of all cropland under cultivation in the United States, are irrigated.[52] This figure represents a substantial increase over the 39 million acres of irrigated cropland reported in the 1969 Census of Agriculture.[53] The bulk of the irrigated acreage—48 million of the total 54 million acres—lies in the seventeen states of the arid West.[54]

The importance of irrigation in the production of food and fiber cannot be overemphasized. Compared to dryland farming, irrigated farming yields a disproportionately large share of total crop production. For the 1969 harvest, an estimated 25 percent of the total cash value represented crops grown on the 10 percent of total harvested acreage that was irrigated.[55] Today, with a greater amount of irrigated cropland, the percentage of total cash value originating from irrigated lands is probably much greater, since average production on irrigated land in the seventeen western states is nearly double production on unirrigated farmland.[56]

The two main sources of irrigation water are surface water from streams and lakes and ground water from wells. To obtain irrigation water, farmers have the choice of diverting surface water, pumping their own water, or paying someone to supply water. Whether irrigation water is diverted from surface sources or pumped from underground reserves significantly affects energy consumption. Diverting surface water uses relatively little energy compared with pumping ground water from deep wells.

Three conditions determine the total energy requirement to pump ground water. First, the average pumping depth, often called the "lift," is directly proportional to the energy needed. In some parts of the West, the lift can range to 300 feet or more.[57] Second, energy consumption depends in part on the efficiency of the pump, in turn determined by the design, condition, and age of the pump as well as by the rate of pumping. Researchers have found that farm pumping plants typically operate at 20 percent below the Nebraska Performance Standard, the generally accepted benchmark of comparison, owing to deficiencies that could be corrected.[58] Third, the energy efficiency of the power unit affects total energy consumption; the conversion of fuel energy into mechanical energy varies from energy-inefficient natural gas engines to relatively energy-efficient electric motors.[59] One estimate puts the energy required to pump an acre-foot of water from an average depth of 200 feet at about 4.5 million BTUs.[60]

Once brought to the surface, irrigation water must be distributed. Surface water irrigation generally requires only small amounts of energy. Sprinkler irrigation, on the other hand, depends on high water pressure, and thus large amounts of energy, to disperse water evenly across a field. Consider, for example, the recent proliferation of huge center-pivot sprinkler systems in the U.S. Midwest.[61] In Nebraska alone, over 1.3 million acres of farmland are now irrigated by these sprinklers. The stunning capacity of a single center-pivot machine to apply water over 133 acres of a 160 acre quarter section depends on massive expenditures of energy. In Nebraska, center-pivot systems consume the equivalent of fifty gallons of diesel fuel to apply about twenty-two acre-inches of water a year—ten times the amount of fuel used to till, plant, cultivate, and harvest an acre of corn in the same state.[62]

Dan Dvoskin and his colleagues at Iowa State University have painstakingly calculated the energy requirements to pump and distribute irrigation water in the seventeen western states in 1975. They report that, on the average, irrigation consumed about 5.2 million BTUs an acre; the energy used to irrigate the 48 million acres in these seventeen states alone was 250 trillion BTUs.[63] For the entire United States, of course, the figure is considerably larger, since the energy needs for the additional 6 million acres of irrigated cropland must be added to this total. U.S. Department of Agriculture economist Gordon Sloggett recently reported that the energy used for irrigation in the United States in 1974 was 260 trillion BTUs, well in keeping with Dvoskin's calculation.[64] These figures are nearly double the oft-cited estimate of 139 trillion BTUs calculated by John and Carol Steinhart for 1970, reflecting increases in both energy consumption per acre and total irrigated acreage.[65]

Although the number of irrigated acres has been rising rapidly, Dvoskin and his co-workers expect a leveling off after 1980, owing both to the expected completion of most surface storage irrigation development projects now underway and to the depletion of ground water in many areas. Nevertheless, development in some areas may continue the demand for ever more energy to power irrigation pumps and sprinklers. In Nebraska, the present 5.6 million acres of irrigated land is being supplemented at the rate of about 300,000 acres a year, with no apparent slacking off in sight.[66]

In many states, farmers and policymakers alike are concerned that pumping for irrigation water is placing a strain on underground water reserves. If the underground water table is to be maintained at a constant level, water can be pumped to the surface only as quickly as it is replenished at the source. Farmers must take care that they do not withdraw water too fast. An average center-pivot system in Nebraska, for example, pumps water at a rate of 900 gallons per minute, consuming enough water to supply the needs of a city of 10,000 people.[67] If water is withdrawn faster than it is replenished, deeper wells must be dug, and that requires more energy and more money for pumping. If the level of ground water reserves continues to decline in an area or if energy sources become scarcer and costlier, irrigation of cropland in the area may become unprofitable or even impossible. Ultimately, these problems will be translated into higher retail food prices or reduced production, or both.

Suggested Changes

Experts agree that many measures can be taken to increase the energy efficiency and decrease the cost of irrigation. The potential savings can be as much as 50 percent of the energy now consumed.[68]

What these changes boil down to are improved management of irrigation water application and improved energy efficiency of irrigation pumping equipment.[69]

Matching the amount of irrigation water to the biological needs of each particular crop can help. Most farmers rely on such surface systems as flood or furrow irrigation, which wastes water and energy and can damage the environment. To make sure that crops at the far end of the field get enough water, farmers continue the flow of surface water for some time after water reaches the far end. Using this system, only 60 percent of the water applied is actually absorbed by the crops.[70] The remainder is lost via evaporation, deep percolation, or surface run-off. In addition, overapplication of irrigation water aggravates problems of water pollution, as silt, salts, and pesticide and fertilizer residues foul the run-off water.

To improve the efficiency with which water and energy are used, farmers can consider the following techniques found in Table 3-2 and the remainder of this section.

Install Surface Run-off Reuse Systems. The effectiveness of surface irrigation systems in getting water to the plants can be enhanced if run-off is directed into collection ponds, from which the water can be reapplied. In this way, average irrigation efficiency can be increased to 85 percent, comparing favorably with that of high frequency sprinkler systems.[71] Construction of run-off reuse systems to recycle water is relatively cheap, and the savings in water and energy are approximately 35 percent.[72]

The major weakness of a run-off system is a potential problem of reduced crop yield due to pollution. Run-off reuse systems may contain unfavorable concentrations of salts and fertilizer and pesticide residues.

Install Sprinkler Systems. Sprinkler irrigation systems that apply water at frequent intervals deliver a greater percentage of their water to the crops than do flood and furrow irrigation ditches. On the average, 75 percent of water applied by sprinklers is absorbed by crops.[73] Indeed, some systems are even more effective; center-pivot sprinklers provide an irrigation efficiency of 85 percent.[74] By distributing irrigation water evenly over the crops with a minimum of run-off, sprinkler systems use less water to irrigate a field than would a surface system, and this results in a savings of energy. In addition, water pollution is diminished, nitrogen fertilizer is saved, and labor requirements are substantially reduced.

Despite these benefits, sprinkler systems have their costs. Large amounts of energy are required to pressurize the sprinkler mecha-

Table 3-2. Energy Savings in Irrigation

- Consider whether flooding in rows or checks, sprinkler, or drip method best fits your condition 15–20%

- Trickle irrigation can conserve considerable energy 4– 8%

- Operate electric motor-driven irrigation wells during off-peak hours 3– 5%

- Install a re-use (tail-water) system on all surface irrigation systems 3– 5%

- Adjust or re-engineer irrigation pumping plants to meet recommended performance standards 2– 5%

- In hot dry areas, consider irrigating at night to reduce evaporation 3– 4%

- When using gated pipe and siphon tubes with irrigation, consider installing automated gated pipe systems with a re-use system 3– 4%

- Know how plants respond best to different soil moisture conditions 3– 4%

- Feel and appearance of soil moisture and rooting conditions can be judged by using a soil spade and auger. Tensiometers and blocks work well in certain soils and crops 3– 4%

- The water budget approach is also used as an additional aid in scheduling irrigation application 3– 4%

- Hold pipe line and fitting losses to a minimum 2– 3%

- Irrigation practices for different crops vary. Check with farm advisors for pre-season and main season, and cut off irrigation practices 2– 3%

- Use aids to schedule correct months for irrigation water . . . 2– 3%

- With manually operated surface irrigation systems, use a water meter or some other method of measuring water . . 1– 3%

- Plan to conserve water use. Less water pumped means that less energy will be used 4– 5%

- Efficiency of any pumping unit should be tested occasionally to determine excessive wear and energy usage. Pumping tests will assist in determining if changes are needed. (Most power suppliers offer a test pumping service) 1– 3%

- Consider land-leveling needs. Well-prepared land surfaces are particularly important in increasing water use efficiencies for surface irrigation 4– 5%

- Plan land use in cropping systems to fit soil and water conditions. Certain crops utilize water more efficiently than others in various soils 3– 4%

- Check on deep percolation losses in conveyance ditches, in storage areas, and in field application 3– 5%

Estimated percentage savings based on total expenditure of energy use. Actual savings will vary depending on specific applications, but should fall in these ranges. These savings are not necessarily additive.

Source: G.B. Taylor, "Agricultural Energy Use," in *Efficient Electricity Use*, ed. Craig B. Smith, prepared for the Electric Power Research Institute (New York: Pergamon Press, 1976), p. 413.

nism for effective distribution compared with surface irrigation systems that require relatively little energy for distribution. Furthermore, sprinkler irrigation systems call for heavy capital investments.

Install Trickle Irrigation Systems.   The trickle irrigation technique works at an average irrigation efficiency of 90 percent.[75] Consequently, the amount of water applied to a crop can be cut down considerably, allowing farmers to reduce their consumption of water and energy. Trickle irrigation minimizes run-off, and thus, fertilizer application can be decreased. In addition, applying irrigation water only to the crop stifles weed growth and less herbicides are used. Finally, farmers can sometimes apply brackish water, putting otherwise unusable water to productive use.

Despite these many advantages, trickle irrigation has some serious limitations that may restrict its usefulness to special applications. First, a trickle irrigation system is very expensive to install and therefore economically feasible only for high value specialty crops. Farmers growing citrus and other orchard fruits may find trickle irrigation cost-effective, but farmers growing most grain crops will not. An elaborate technology governs the use of trickle irrigation, and equipment tends to clog with silt and iron from the water. The necessary filtering devices are costly and require high pumping pressures, which decreases the energy efficiency of this method somewhat. Finally, some farmers have experienced temporary reductions in yield as crops react to the new watering system.

Despite all these limitations, CAST estimates that use of trickle irrigation could replace some energy-inefficient irrigation systems to save the equivalent of 27 million gallons of gasoline, or 3.4 trillion BTUs, every year.[76]

Improve Irrigation Scheduling.   Through improved irrigation scheduling, energy and water consumption can be reduced substantially. For example, soil moisture sensors can be used to tell farmers the precise water needs of a crop. Farmers could then apply programmed soil moisture depletion techniques to maximize the effective use of stored soil moisture and rainfall. Under this scheme, farmers irrigate only when the need has been scientifically verified. These techniques of improving irrigation scheduling could save 15 to 35 percent of the present amount of water pumped.[77]

Schedule Electricity Demand to Avoid Peak Generating Periods. By scheduling their electricity use outside the peak generating periods, farmers and farm electric cooperatives can reduce their electricity bills.[78] If farmers spread out their demand over off-peak hours, they can avoid high peak demand charges, since utilities generate electricity less expensively to meet baseload demand than to meet peaking demand. Reducing peak demand also lowers energy use, because electricity is generated more energy-efficiently to meet baseload demand than it is to meet peaking demand.

Reduce Pumping Pressure of Sprinkler Irrigation Systems. According to one report, lowering the pumping pressure of an average system by twenty pounds per square inch will reduce energy consumption by 13 percent.[79] This technique is useful for farmers who use systems that do not require high pumping pressures.

Improve Irrigation Pump Efficiency. The previously discussed water management techniques can conserve substantial amounts of energy. These energy savings are small, however, compared to the energy savings that can accrue to improvements in pumping plant efficiency. Recent tests have found that the average irrigation pump operates about 20 percent below the Nebraska Performance Standard, the generally accepted standard of irrigation pumping plant performance.[80] Certainly, if all irrigation pumps could be made to meet the Nebraska Performance Standard, the energy savings would be tremendous. A study of energy use in Texas agriculture concludes that realistic improvements in pumping plant performance could save almost 39 trillion BTUs a year in Texas alone, but the cost would be $116 million.[81]

Using the Steinharts' conservative estimate that farmers use 139 trillion BTUs to irrigate their land, CAST estimates that the equivalent of 250 million gallons of gasoline, more than 31 trillion BTUs, could be saved each year through pumping plant improvements.[82] The current, higher estimates of total energy consumption for irrigation would yield correspondingly greater figures for these potential savings.

Beyond the direct savings of water and energy due to improved irrigation techniques, additional energy can be conserved. Run-off water from excessive irrigation percolates deep into soil, leaching away expensive and energy-intensive nitrogen fertilizer. One study suggests that careful scheduling and application of irrigation water can save an average of fifty pounds of nitrogen an acre.[83]

Strategies for Energy Conservation
States can adopt a number of strategies to implement the energy-saving techniques just discussed and thus increase the efficiency of

energy use in irrigation. A checklist of some of these strategies follows:

Fund Educational and Advisory Services. States can tap several sources of expertise to provide farmers with information on how to minimize their energy and water consumption for irrigation. States should assess carefully the need for agricultural extension classes, irrigation advisors, and demonstration projects. Some states already devote plenty of time and resources to such programs. Other states may find that additional funding is needed to supplement educational programs through their agricultural extension services and departments of agriculture. In addition, hiring irrigation specialists to advise farmers can yield handsome dividends in water and energy savings.

States should take steps to see that irrigation pumping plants are upgraded, the one improvement with the greatest potential to reduce energy consumption for irrigation. Inspection, adjustment, and repair of irrigation pumps could be done routinely by state agriculture department personnel or by outside field advisors and service technicians. States should probably charge a fee to cover the cost of this service. To encourage farmers to use the service, states should demonstrate to farmers the cost-effectiveness of improving the energy efficiency of their pumping plants to meet the Nebraska Performance Standard.

Finally, states can provide irrigation technicians to help farmers improve their scheduling practices. Scientific measurements of the soil moisture profile, along with the latest information about such water management techniques as programmed soil moisture depletion, can help farmers eliminate unnecessary irrigation. State agriculture specialists or outside irrigation technicians can plot the irrigation needs of farmers precisely, with resulting savings in irrigation water, money, and energy. Again, states should probably charge a reasonable fee to cover the cost of irrigation scheduling advice.

Fund Research and Development Programs. The National Association of State Universities and Land-Grant Colleges, the U.S. Department of Agriculture, and Congress have recently concluded that much more research is needed on irrigation.[84] Among the irrigation topics identified as deserving attention are improved water distribution methods, water management techniques, and reuse of run-off water.

Research is also needed to develop irrigation pumping plants that do not rely on nonrenewable fuels. The development of solar irrigation pumps is already underway in New Mexico. This project, undertaken by Sandia Laboratories and New Mexico State University, has received $500,000 in funding from the New Mexico Energy Resources Board and other sources.[85]

Prohibit Wasteful Irrigation Practices. Water is an essential, finite resource, and states must take steps to eliminate waste by all water users. In 1975, the Nebraska Ground Water Management Act became law, providing a legal and institutional framework for future management and control of state ground water used for irrigation.[86] The act instructs each of Nebraska's twenty-four resource districts to "adopt, following public hearing, rules and regulations necessary to prohibit surface runoff of water derived from ground water irrigation."[87] Properly administered and enforced, the Nebraska law could help conserve both ground water and energy now lost when substantial run-off occurs.

Kansas, another state that relies heavily on ground water for irrigation, agrees that steps must be taken to conserve ground water reserves. A legislative Special Committee on Energy and Natural Resources recommended in 1976 that ground water management districts require the elimination of all wasteful irrigation practices.[88]

Offer Financial Incentives. Some of the technologies and irrigating techniques suggested above offer farmers the possibility of large savings in energy and water, but not without large capital investments and risks of economic loss. New water management practices may require significant investment in irrigation scheduling programs, soil moisture sensors, water meters, sophisticated sprinkler systems, and irrigation technicians.

Some farmers may be skeptical that the potential savings outweigh the costs of investing in energy-conserving irrigation equipment and techniques. Others may simply lack the necessary capital. States can counter this reluctance to invest through the use of loan programs and tax incentives that encourage farmers to invest in energy-saving irrigation methods.

Two states in the Southwest have already taken initiatives in this direction by enacting legislation specifically geared to encourage investment in solar-powered irrigation systems. New Mexico now allows its farmers an income tax credit of up to $25,000 for investment in solar irrigation pumping systems.[89] And Arizona exempts solar equipment used for irrigation pumping from the use tax and a tax based on sales.[90]

Offering financial incentives to farmers would be politically popular and easily administered, generally calling for no additional bureaucracy. On the other hand, credit subsidy and tax incentive programs are paid for by all taxpayers, while only a small group of farmers benefit. Indeed, financial incentives may have only a small marginal effect and may result in a windfall for those who would have invested in more energy-efficient irrigation equipment anyway. Financial incentives to induce investment should probably be reserved for subsidizing farmers who invest in equipment or methods with a high risk of economic loss, not for bankrolling investment in changes that have clear economic benefits and little economic risk. The investment risk must be appraised before states decide whether to offer tax incentives or low interest loans.

Increase the Price of Water and Energy. For many years both energy and water have been priced below their actual worth as resources, a practice almost guaranteed to result in misallocation. In 1972, before the onset of higher energy prices, researchers at Iowa State University predicted that requiring farmers to pay the full costs of irrigation would reduce water consumption by 50 percent.[91] The National Water Commission recommends that policymakers offer an incentive to water conservation by setting the price of water to cover the social costs—of pollution, erosion, and so on—created by an irrigation project.[92]

Farmers' extreme sensitivity to price increases makes irrigated acreage an early casualty of water or energy price hikes. Researchers at Iowa State University also report that in a simulated energy crisis, cutbacks in irrigation accounted for most of the energy savings.[93] The report forecasts that farmers would respond to a doubling of the energy prices by decreasing their energy consumption by 28 percent.[94] Price regulation is, however, a double-edged sword. On the one hand, the potential for conserving energy in irrigation makes it an inviting target for regulation. But higher irrigation costs could adversely affect food production and retail food prices. States must, therefore, exercise care and restraint in balancing the goal of energy conservation with the goal of food production.

Some states have control over government-supplied irrigation water. Charging more can reduce water consumption, but many states would have to legislate a new charge for irrigation water.[95] In states where charges already have been levied, the price can be reevaluated and, if warranted, raised to reflect the total costs of providing the water.

States can further reduce energy consumption by raising the price of the energy that powers irrigation pumping plants. The Pacific and Mountain states depend on electricity to drive irrigation pumps. In

California, for example, electricity accounts for 82 percent of the energy used in irrigation.[96] The Northern and Southern Plains states rely primarily on natural gas for irrigation. Most natural gas is produced, distributed, and sold in interstate commerce and is regulated by the Federal Energy Regulatory Commission, not by the states. However, some states are large producers of natural gas that is sold within the state, and this intrastate natural gas is subject to state regulation. Through their public utilities commissions, states can raise rates for water and energy with relative administrative ease. The commissions have the institutional framework and personnel to conduct hearings and to issue orders with dispatch.

In addition to raising water and energy prices directly through rate regulation, states can tax these goods. A sales tax collected as a percentage of the sales price of water or energy or an energy tax based on the total amount of energy consumed could be the method chosen.

Allocate Irrigation Water. When farmers fail to consider the biological needs of plants and apply more water than necessary to replenish moisture and to flush salts from roots, the result is water lost to run-off. Energy used to pump and distribute the excess water is lost as well. Water allocation programs could help cut these losses of water and energy.

By definition, a water allocation program limits the amount of water available to irrigators. Water consumption can be monitored and controlled by metering water as it is withdrawn from the source, by limiting the operating time of water pumps, by scheduling irrigation applications to conform to pumping rotations, by setting standards for well spacing and well size, and by various other means. Water allocation strategies, of course, must be designed to provide enough water to maintain desired levels of crop production and farm income while assuring adequate supplies of water for the future. The goal is to balance these needs politically, socially, and economically.

Mandatory restrictions or quotas on irrigation water can probably help save water and energy, but not without problems. First, legal complications accompany any program that tells farmers they cannot use a volume of water previously acquired. In the West, where most irrigated acreage lies, water is acquired as a property right obtained by developing the source and putting the water to a beneficial use. Since the right is subject to continued beneficial use, farmers who reduce the quantity of water diverted for irrigation purposes run the risk of losing the right to the undiverted water. A state that restricts the amount of water available to farmers may be infringing on a vested property right, which might be challenged as a taking of property without due process of law.

In practice, a mandatory limit on the availability of water may have some very impractical consequences if decreased crop production, lower farm profits, or higher food prices result. Crop yield depends upon the availability of sufficient water to plants at critical stages of their development, and a lack of water at that time can spell disaster to crop yields and farm profits.

A third problem with mandatory restrictions is the need to maintain flexibility. Different soils and different crops have different moisture retention capabilities, so that an arbitrary allotment of water could be just right for one farmer but not enough for another. An allocation program must be made flexible enough to accommodate individual needs.

And so it seems that mandatory water allocation programs must be approached with caution, if at all. A better idea might be for farmers to join with state water conservation agencies in founding a voluntary, cooperative program for water conservation. Such a program could lead to substantial savings in water and energy while avoiding the problems of the more coercive mandatory program.

Farmers and local water conservation officials have organized a program of voluntary water allocation in Nebraska and have demonstrated its effectiveness. In 1974, the farmers and ground water conservation districts in five Nebraska counties began the Benedict Project to investigate whether ground water allocation could succeed.[97] Emphasizing current water management technology, the farmers and irrigation specialists found that irrigators could readily adjust to the water use limits set as goals. During the summer of 1975, the project conducted a demonstration that included the use of water metering, irrigation scheduling, soil moisture monitoring, and determination of ground water withdrawal limits in advance of the growing season. The Benedict Project yielded several positive results. First, the goal of reducing ground water use was easily achieved on both the pilot fields and the ordinary farms. Second, a practical system for water allocation was shown to benefit the cooperating farmers who made the system work. Third, farmers succeeded in cutting energy consumption: irrigation pumping plant adjustments and improved water management techniques produced a 10 percent increase in energy efficiency.[98]

Conceived in a spirit of cooperation, water allocation programs can effectively reduce water and energy consumption. Conservation of energy and resources is essential, but it must be achieved in concert with other socially desirable goals—the economic well-being of farmers, full crop production, and reasonable food prices.

Restrict Availability of Energy Resources. Limiting the amount of energy available for irrigation is a step that must be taken cautiously, in order to avoid unsettling economic, social, and political consequences. Energy allocations are, in effect, the same as water allocations, since most irrigation systems depend on energy to pump and distribute the water. Thus, many of the considerations described in the preceding section are applicable here.

Irrigation load management is not a new or an untried concept, and the results of three of these systems show promise. For example, the Custer Public Power District of Broken Bow, Nebraska, requires irrigators to accept interruptible service, which guarantees the irrigator sixteen hours a day of uninterrupted service.[99] Another utility, the Elkhorn Public Power District of Battle Creek, Nebraska, uses a different system. New irrigators are given the option of accepting control either by time switches or radio control.[100] A third utility, the Southwest Public Power District of Palisade, Nebraska, uses a ripple control system. Existing irrigators are given the choice of whether to join the program; new customers must accept load management in order to receive service.[101] These three systems have demonstrated that a properly designed and administered system of limited energy availability can have a positive effect, saving farmers money and energy. Care must be taken, of course, to avoid excessive energy restrictions that can reduce crop yield below economic savings.

CROP DRYING

To keep harvested crops safe from the deteriorating effects of heat buildup, mold formation, and insect infestation, farmers must reduce the moisture content of crops below some critical level. The maximum acceptable moisture level varies, of course, with different crops. For example, field corn is generally harvested when the moisture content is between 25 and 30 percent, but must be dried to a moisture level of approximately 15 percent if it is to be safely stored for later use as animal feed.[102]

Traditionally, farmers dried their crops by allowing the sun and wind to evaporate the excess water. This method was, however, frequently accompanied by reduced crop yields due to weathering, storms, and predator damage. Indeed, prior to 1945, the drying of crops with artificially heated air was an experimental novelty, except for tobacco.

A typical artificial crop dryer uses convection type heaters that burn some type of fuel, usually liquefied petroleum (LP) gas or natural gas. Drying occurs when the heated air, which can hold more moisture than cool air, is forced through grain in the drying bin by an electrical fan or propellor. Such convection type, heated air dryers are roughly 50 percent efficient in utilizing fuel to evaporate moisture.[103] To get an idea of the amount of fuel consumed and the heat produced by these crop dryers, consider that a typical three bedroom house of 1,200 square feet may have a furnace that generates 150,000 BTUs and uses the equivalent of one and a half gallons of LP gas an hour. Small crop dryers used in deep bed grain drying produce at least 1 million BTUs while consuming about ten gallons of LP gas an hour; high speed dryers used on farms and in elevators typically generate 3 million to 6 million BTUs by burning thirty to sixty gallons of LP gas an hour; and dryers at the large terminal elevators may have far higher fuel requirements.[104]

In the last several years, there has been a dramatic increase in the heated air drying of farm products, especially corn. In fact, crop drying has become so common that American farmers in 1973 burned the equivalent of 1.3 billion gallons of LP gas, or 120 trillion BTUs, in drying corn, tobacco, soybeans, rice, peanuts, sorghum, and other crops.[105] The drying of the corn crop alone consumed 56 trillion BTUs, more than thirty times the energy used to dry corn in 1945.[106]

This sizable increase is the result of two major changes in corn production. First, most farmers have switched from harvesting whole ears of corn and drying them naturally to the more economical method of shelling corn in the field and drying the kernels artificially. The conversion to field shelling has been quite rapid throughout the country, but is most noticeable in the Corn Belt, where the total acreage that was field-shelled increased from 15 percent in 1960 to 77 percent in 1972.[107] Second, many farmers are harvesting their crops earlier to reduce the risk of bad weather loss, even though an earlier harvest means that the crop will have a higher moisture content and will require additional drying before it can be safely stored. Artificial drying has increased accordingly, and in 1973 about 70 percent of the harvested corn crop was dried this way.[108]

Suggested Changes

The relatively recent development of heated air drying suggests that the level of artificial drying can be reduced and, consequently, that a lot of fuel and energy can be conserved. Although some crops, like rice and tobacco, resist efforts to reduce or eliminate drying, the amount of energy consumed by most other crops can be minimized. Corn appears to be the crop with the most potential for reduced energy use, and several approaches have been proposed.[109]

Return to Older Drying Methods. One technique for reducing fuel and energy requirements is to return to the methods that were formerly used to harvest and condition corn. Farmers could harvest whole ears of corn and allow them to air dry in bins naturally, rather than shelling the corn in the field and drying it artificially. Farmers could also allow the corn to stand in the field until it had dried to a lower moisture content before harvesting. Neither of these suggested changes is acceptable when considered alone as an alternative to the current method of using heated air drying systems.

The advantage of either of these methods is, of course, the elimination of the need for LP gas or natural gas to fuel crop dryers. Unfortunately, returning to the older drying methods has many potential costs. Relying on field drying alone, for example, creates a substantial risk of reduced crop yield due to delays in harvest and adverse weather conditions. It has been estimated that the reduction in yield could be as much as 10 percent if the weather conditions were especially bad.[110] There is the potential for total crop loss if the weather is severe and farmers do not have livestock to glean the remaining crop from the field. Some agricultural researchers believe that field drying is a long-term answer only if fuel becomes totally unavailable for drying.

Changing back to the method of harvesting whole ears of corn is also fraught with problems. The equipment needed to harvest whole ears of corn is presently available for only one-half of the corn crop. A reversion to this method would require the production of completely different harvesting, handling, storage, and processing equipment. Producing this equipment would necessitate tremendous capital investment in manufacturing facilities and steel products, consume great amounts of energy, and take about five years.

Even then, the new system would still encounter harvesting difficulties. Since ears of corn occupy twice as much space as the same amount of shelled corn grain, and because ear corn does not flow freely, twice as many vehicles (or vehicles that move twice as fast) would be needed. In either case, these vehicles would consume more fuel and energy than the vehicles that are now used to harvest the corn. It has been estimated that a change to the older method would reduce net harvested yield by 10 to 15 percent, as timeliness in harvest is lost or as shorter season varieties are adopted.[111] Storing ear corn in open air facilities subjects the crop to greater damage by rodents, birds, and contaminants in the air. This lowers both the quantity and the quality of the yield, increasing the amount of energy consumed for each unit of output.

Air Dry Crops.   A more practical way to increase the energy efficiency of crop drying systems is to utilize methods that reduce fuel requirements without sacrificing crop quality. Farmers can dry their crops in several ways to decrease the fuel consumption in this operation. For example, they could dry their crops with unheated air, a change that could be accomplished easily in a short time. This method uses air at its ambient temperature, circulating it through the storage facility by an electric fan. Although this system requires more electricity to power the circulating fans than do the heated air methods, neither LP gas nor natural gas is used, and conversion to this method would result in a large, but unquantified, net reduction in energy consumption. In addition, this method does not reduce crop yield.

On the other hand, drying a crop with unheated air takes longer than using heated air dryers. Converting to this system would require major new construction of drying and storage facilities, necessitating increased manufacture of energy-intensive steel products. Although this method could be used to reduce energy consumption, several alternative heated air drying techniques hold more promise.

Adopt Dryeration Method.   Even if farmers and grain elevator operators continue to dry crops with artificially heated air, which seems to be a fair assumption, significant differences exist in the energy efficiency of available systems. The widespread adoption of the dryeration technique, a process that is used in conjunction with a regular heated air system, may save a lot of fuel.

Dryeration works this way. After a normal period of drying to remove half the moisture, the grain is removed hot from the drying bin, transferred to a storage bin where it remains for four to twelve hours without any air flow, and then cooled slowly with aeration. The final drying and slow cooling result in the removal of the remaining excess moisture. This energy-efficient and effective technique could be quickly applied to many existing farm and elevator drying systems. Dryeration has several clear advantages: it increases the drying capacity of a system by 50 to 60 percent, decreases the fuel requirements by 20 to 30 percent, and improves the marketability of the dried crop as compared with other high speed drying techniques.[112] It is estimated that the annual energy savings from greater implementation of the dryeration process could be about 3.7 trillion BTUs.[113] This represents almost 7 percent of the fuel and energy consumed in drying the corn crop or about 3 percent of the fuel and energy used to dry all crops.

Of course, the dryeration technique is not without costs. Implementing dryeration requires farmers to add one or two dryeration cooling bins to their present drying system or to adapt some existing storage bins to such use, a change that would require a substantial capital investment. This added expense yields an additional bonus, however, as these bins may also be used for regular grain storage at the end of the drying season, thereby partially offsetting the price. In summary, dryeration appears to be an alternative with great potential for energy conservation, and farmers should be encouraged to use this technique.

Convert to Solar Crop Drying. Farmers can substitute solar energy for the LP gas and natural gas that are now used to dry crops. In several respects, the application of solar energy to crop drying is an ideal use of the sun's heat. Harvested corn can be dried to a safe moisture level by a crop drying system that produces only low temperature rises, which is exactly what solar drying systems do. Although low temperature drying systems require a longer time to dry the crop completely, they minimize kernel damage and yield a higher quality final product than high speed, high temperature dryers. In addition, harvested shelled corn can be dried over a relatively long period of time without continuous heat flow, as long as the corn is kept well ventilated. Thus, the use of solar energy to dry crops can work even if the sun does not shine for several days and even if the system does not have a capacity for heat storage.

The chief advantage of solar drying units is that energy costs, both in dollars and in gallons of refined petroleum products, can be reduced drastically. Solar energy, according to an estimate by ERDA, can be substituted for at least 500 million gallons of LP gas and natural gas consumed annually in corn drying, representing an energy saving of over 46 trillion BTUs a year.[114]

Experts now disagree about the current economics of solar grain drying systems, but everyone concurs that solar energy looks better every time fuel prices go up. Agricultural researchers who find that solar grain drying is marginal from an economic standpoint concede that the economics could quickly change if tax breaks were allowed or if technical improvements reduced the cost of solar collectors.[115] Some farmers and agricultural researchers have found that solar grain drying systems are a good investment right now. One Wisconsin farmer who installed a solar grain dryer reported a one-third reduction in the cost of drying corn in the first year; at the end of the second year, it had paid back almost three-fifths of its cost.[116] Gene Shove, an agricultural engineer at the University of Illinois, concludes that solar collectors and grain drying systems will easily pay for themselves over three to five years.[117]

On the other hand, low temperature solar grain drying systems have some limitations that may retard their adoption. First, solar grain drying, like any low temperature drying system, is dependent on the weather. Successful drying without spoilage works only with certain minimum ambient air temperatures and maximum initial grain moisture content. To some extent, solar drying systems are experimental, with research currently underway at twelve agricultural experiment stations.[118] In addition, the economics of using solar drying systems are still open to question, based on the current prices of fuel and of solar collector equipment.

One idea that has garnered interest is to combine the best qualities of high temperature dryeration with low temperature solar drying. Under this regime, most of the moisture in the corn would initially be removed by the dryeration process. The grain would then be transferred to storage in a low temperature solar drying system where the corn would be dried slowly over time. Because high temperature drying is most efficient with high moisture corn and low temperature drying is most efficient with low moisture corn, this system would combine the best parts of both.

Use High Moisture Corn. Corn can be taken directly from the field and stored in its high moisture condition without the necessity of drying, if it is stored in an airtight container or if it is treated with chemical preservatives to inhibit mold growth.

Using airtight containers and high moisture corn preserved with organic acids can conserve all the fuel that would have been burned to dry that corn. Using high moisture corn yields an additional bonus: cattle are able to utilize high moisture corn more effectively, gaining more weight per unit of feed on wet corn than they do on dry corn.[119] Livestock are also able to convert some of the preservative chemicals in the feed into weight gain, thereby lowering the net energy cost of using this method.

Unfortunately, there are at least four major problems impeding the rapid conversion to wet grain storage. First, when organic acids are used to treat corn, its use is restricted to livestock feed only. Indeed, there are confines on that use as well, because hogs, unlike cattle, gain less weight per unit of feed on a diet of wet corn than they do on dry corn.[120] Another disadvantage is that the amount of petroleum feedstock required to synthesize the organic acid costs as much as the LP gas and natural gas required to dry an equivalent amount of corn in a heated air dryer. When processing and distribution costs are included, the total dollar cost of the wet grain storage method exceeds the dollar cost of heated air drying. Third, the acid used to treat the corn corrodes most materials from which grain storage facilities are constructed. Steel and concrete are especially vulnerable. Finally, if this method of grain preservation were widely used, a great many on-farm storage facilities would have to be built.

Table 3—3, which is taken from a report prepared for the Electric Power Research Institute, offers suggestions for energy conservation similar to those discussed previously and estimates the percentage of energy savings possible if these changes were implemented.

Strategies for Change

The preceding sections have introduced techniques to decrease the amount of energy consumed in drying crops. Agricultural policymakers must, of course, devise programs that selectively combine the advantages of different methods while minimizing their shortcomings. The most feasible alternatives appear to be increased use of dryeration, solar grain drying, and high moisture corn. Suggested strategies include:

Provide Information About Various Crop-Conditioning Methods. Farmers who are interested in minimizing the operating costs of production will be receptive to suggested changes in crop-drying meth-

Table 3—3. Energy Savings in Crop Drying

• Harvest and handle as high-moisture shelled corn if an outlet is available	20–25%
• Harvest as corn silage if an outlet is available	10–20%
• Use natural air drying and low temperature drying as much as possible	10–15%
• Delay harvest as long as it is practical to dry on the stalk . .	5–10%
• Consider storing this year's harvest on the farm	5–10%

Estimated percentage savings based on total expenditure of energy use. Actual savings will vary depending on specific applications, but should fall in these ranges. These savings are not necessarily additive.

Source: G.B. Taylor, "Agricultural Energy Use," in *Efficient Electricity Use*, ed. Craig B. Smith, prepared for the Electric Power Research Institute (New York: Pergamon Press, 1976), p. 403.

ods, if they can be convinced that the new system is both reliable and economically beneficial. To help farmers make a rational decision about whether to adopt a new method of crop drying, states should offer the most recent information that explains the reliability and economic benefits of each alternative. The outreach programs of the Cooperative Extension Service are the ideal way for states to help their farmers know whether they can benefit from the use of alternative crop-conditioning methods, especially solar energy.[121] States can choose from many strategies, including the use of classes, workshops, written publications, and individual consultations.[122]

Another educational step that states can take to help inform farmers about the energy efficiency and lifetime operating costs of particular crop dryers is to require fuel efficiency labeling on crop dryers. Such a program could significantly influence an individual farmer's purchasing decision in the same way that fuel efficiency labeling now influences an individual's selection of a passenger automobile.

Offer Inspection and Adjustment Advisory Services. To help farmers maintain their crop-drying equipment at peak operating efficiency, states could offer the services of trained technicians to inspect and adjust grain dryers. Both farmers and society would benefit from such a program: proper adjustments will help reduce a farmer's costs of crop drying, and fuel will be conserved for use elsewhere.

The field advisors for this program could be either state-employed experts or independent technicians hired to inspect and adjust crop dryers. Farmers should be charged a fee for these services to cover the costs, but states could encourage participation by subsidizing the services to some extent.

Provide Loans for Investment in Energy-Saving Equipment. Before farmers will invest in the equipment or chemicals necessary to implement a particular energy-efficient system of crop conditioning, they must be convinced that it will be reliable and that its costs will not outweigh its benefits. Dryeration, for example, is a reliable, relatively risk-free method of drying crops safely and economically. But conditioning crops with solar drying equipment or chemical preservatives presents a riskier picture. These last two methods have great potential for reducing energy use, but both have certain limitations that may cause cautious farmers to pause before committing large sums of money.

Even if farmers can be convinced that these new methods for crop conditioning hold the promise of improved energy efficiency and reduced operating costs, they may still hesitate to invest large sums of money in these methods because of the initial high cost of equipment. For example, farmers who want to adopt solar crop-drying methods will find that solar collectors and dryers can vary tremendously in price.[123]

States can encourage farmers to invest in such energy-efficient capital equipment as solar crop dryers or new storage facilities by providing credit through state loan programs. For example, a state could provide direct state loans to farmers from the state treasury at favorable interest rates; it could provide loan guarantees to private lenders who would make the loans to farmers; or it could reduce the effective cost of investing in new equipment by offering interest subsidies on loans for energy-saving crop-conditioning systems. To stretch the budget as far as possible, states should probably tie loan eligibility to both energy efficiency and potential risk of economic loss. Thus, states could offer loans for the purchase of solar drying equipment or high moisture grain storage bins or equipment that harvests ears of corn, but not for an energy-efficient method such as dryeration that presents little risk of economic loss.

Provide Tax Incentives for Investment in Energy-Saving Equipment. Tax incentives are a proven way to influence behavior, and states could encourage investment in energy-conserving crop-conditioning equipment by allowing tax breaks. Tax incentives could take the form of investment tax credits, tax exemptions, or tax deductions for the purchase of energy-saving equipment that presents a significant economic risk. Thus, states could consider allowing tax breaks to farmers who purchase solar drying equipment or who retrofit existing storage facilities to new energy-efficient crop-drying uses.

Offer Crop Protection Insurance to Farmers Who Use High Risk, Energy-Saving Techniques. The use of solar collectors, unheated air, or natural air drying of whole ears of corn holds the promise of great energy savings. Adopting these techniques, however, may increase the risk of economic loss compared with conventional drying systems, and farmers, quite understandably, will avoid this risk. Farmers who choose to adopt energy-saving techniques are helping society by freeing fuel for other uses, and states can return the favor by offering some form of crop protection insurance for farmers who utilize energy-saving but risky methods. States could subsidize farmers who use alternative, risky techniques either by guaranteeing them a certain price for their harvest or by subsidizing the cost of crop protection insurance premiums.

Subsidize the Purchase of Energy-Efficient Equipment. If using a new energy-saving technology creates a substantial economic risk or requires a large initial capital investment, states can take steps to encourage manufacturers and farmers to make and use such equipment. For example, farmers who use solar drying equipment may run the risk of crop damage if the crop is harvested at a high moisture content and the weather is especially adverse. Understandably, farmers may not want to lock themselves into a position of depending entirely on such a system.

To help farmers convert to a system that combines the best features of both conventional and solar dryers, states could purchase solar drying equipment from the manufacturer at the market price and sell the equipment to farmers at a discount. Such a strategy encourages the manufacturer of solar equipment by providing a reliable market, and it encourages farmers by making the equipment less expensive. Each state must, of course, decide as a matter of policy whether the benefits of reduced fuel consumption exceed the cost of such a subsidy.

SPECIALIZED OPERATIONS

Most Americans like to eat fresh fruits and vegetables throughout the year, and many people are fond of buying ornamental cut flowers and decorative nursery plants, even in the dead of winter. Consumer demand for these products creates a substantial demand for fuel and energy, since the climate in most parts of the country does not normally permit the year-round cultivation of these crops. To meet the consumer demand for such high value specialty crops as tomatoes and citrus fruits, farmers have adopted energy-intensive cropping techniques, including greenhouse cultivation and the use of frost protection devices. The amount of energy used for these operations is small compared to the energy consumed by farm machinery, irrigation pumps, and crop dryers, but there is room for significant fuel savings.

Greenhouses

In 1970, commercial greenhouse growers cultivated over 275 million square feet or about 6,300 acres. Although this is a trifling amount compared to the millions of acres that are planted to corn, wheat, and soybeans, greenhouse heating and ventilation are quite energy-intensive, consuming the equivalent of over 40 trillion BTUs annually.[124] During 1969, farmers in California alone cultivated over 70 million square feet of greenhouse space and consumed almost 8 trillion BTUs.[125] Moreover, energy consumption is rapidly expanding as the acreage of greenhouse cultivation increases. One report, based on data from 1972, estimated that California would experience greenhouse sprawl at a rate of about 9 percent a year and that energy consumption would grow at the same rate.[126]

Because fuel costs constitute a large portion of the budget in a greenhouse operation, farmers have an economic motivation to take energy-conserving measures. Fortunately, many simple housekeeping improvements can be used to conserve energy at minimal cost. Basically, farmers can increase energy efficiency by adding insulation, setting thermostats for optimum growth, and maintaining equipment.[127] It has been suggested that farmers take the following steps to improve fuel efficiency:

- Adjust burners and stokers of boiler for peak efficiency,
- Repair greenhouse openings to stop leaks,
- Paint greenhouses white to reflect unwanted heat,
- Check thermostats for proper operation regularly,
- Insulate by putting plastic sheeting on side walls and in walls of greenhouse,
- Maintain the heating pipe system to stop leaks,
- Maintain the automatic valves in the heating system,
- Use reflecting tarpaper behind heating pipes to spread heat into the greenhouse,
- Maintain insulation on the pipes,
- Use black cloth on the greenhouse structure at night to reduce radiation loss to the atmosphere, and
- Do not steam sterilize more than necessary.[128]

CAST estimates that following these simple maintenance and operating procedures could save 20 to 25 percent of the fuel used, representing a savings of 5 trillion BTUs.[129]

Table 3-4, taken from a report prepared for the Electric Power Research Institute, lists some of the energy-conserving suggestions and estimates the potential percentage of energy savings that would accompany their implementation.

Table 3-4. How to Save Energy in Greenhouses

• Adjust burners and stokers	10–15%
• Repair greenhouse openings	8–10%
• Paint greenhouses white	5– 8%
• Be sure temperature instruments are working correctly . . .	4– 8%
• Put plastic on side walls and in walls of greenhouse	4– 6%
• Do not steam sterilize more than necessary	2– 5%

Estimated percentage savings based on total expenditure of energy use. Actual savings will vary depending on specific applications, but should fall in these ranges. These savings are not necessarily additive.
Source: G.B. Taylor, "Agricultural Energy Use," in *Efficient Electricity Use*, ed. Craig B. Smith, prepared for the Electric Power Research Institute (New York: Pergamon Press, 1976), p. 402.

Another suggestion, which is more ambitious than the housekeeping measures described above, is to utilize the waste heat generated by electric power plants for heating and cooling greenhouses. The generation of electricity always produces heat, an unwanted by-product that is usually considered to be waste. This heat, however, is a valuable resource that can be used to increase food production and to reduce energy consumption, thereby raising the energy efficiency of agriculture. Several investigators have studied how power plant cooling water can be reused. They report that farmers, especially greenhouse growers, can utilize low temperature waste heat from power plants without reducing electrical energy generation. The reports show that warming and cooling greenhouses with waste heat can improve crop growth and yield, can conserve fossil fuel resources, and can save money for both farmers and the electric utility.[130] In addition, utilizing waste heat from electric generating plants can reduce the impact of thermal effluents on the local ecology. The major constraint to the quick adoption of this technology is the price of putting the system into operation; currently such systems are too costly to make investment economical.

Another innovative idea that is presently undergoing intensive study by the federal government is the use of solar energy to heat and cool greenhouses.[131] The design and construction of solar greenhouses represent an application of high technology to energy conservation. Many people already use solar-heated, energy-conserving greenhouses, and they are becoming increasingly attractive for use on farms, in suburban gardens, and on city center rooftops.[132]

Frost Protection
Such high value specialty crops as citrus fruits, deciduous fruits, grapes, and nuts are especially sensitive to sudden chilling frosts.

Orchard growers who produce these high value crops use energy-intensive wind machines, heaters, and sprinklers to protect their crops against damaging frost. Of course, one of the reasons that frost protection methods are necessary in the first place is that the production of frost-sensitive crops has been expanded to marginal areas that experience a greater degree of adverse weather than the crops can naturally withstand.

Because frost protection methods are used only on a few high value crops and because the use of these methods is intermittent, the absolute level of energy consumption is minute compared to the energy consumed in machinery operation, irrigation, and crop drying. Nevertheless, in some states the use of fuel for frost protection merits attention as one area in which energy can be conserved.[133] California, for example, is one of the leading states in producing citrus and other fruits. In 1972, the energy expended to power frost protection devices was over 9 trillion BTUs.[134]

Energy expenditures could be reduced if farmers replaced heaters with sprinklers and wind machines and shifted from gasoline-powered engines to electric motors. CAST estimates that both these changes could save the equivalent of 500 billion BTUs annually.[135] Further strategies might be to:

Provide Farmers with Information About Energy Conservation. States should probably first examine whether their information transfer programs are successfully disseminating helpful energy conservation ideas to farmers. If not, states could increase funding to their land-grant colleges and cooperative extension service specialists. Additional funding will permit extension service specialists and county agents to conduct more classes, personal consultations, and demonstration projects. In addition, an effort can be made to distribute publications that explain how energy can be conserved in commercial greenhouse cultivation or frost protection.

Fund Additional Research in Energy-Saving Frost Protection Techniques and Greenhouse Cultivation Methods. The use of solar equipment or waste heat are relatively new technologies for heating and cooling greenhouses and for protecting crops from frosts. Although research is now underway, much more can be done to determine how these technologies can be used to lower the cost of their installation. States could provide additional research funding in conjunction with USDA and ERDA, both of which have undertaken research programs or supported projects to study the use of new technologies.[136]

Provide Advisory Services to Improve Operation and Maintenance of Greenhouses and Frost Protection Machinery. To help maintain a peak level of fuel efficiency in greenhouses and frost protection machinery, states could offer the services of trained mechanics to inspect and adjust the equipment for optimal performance. Discovering energy waste through energy audits, inspections, and adjustments can help farmers quantify their potential energy and cost savings. States should require farmers to pay for at least some of the cost of providing these services. A reasonable charge is unlikely to deter farmers when the savings in fuel will more than offset the cost of the inspection and adjustment.

Provide Loans for Investment in Energy-Saving Equipment. Although some of the housekeeping suggestions can be implemented cheaply, other changes are neither simple nor inexpensive. Installing solar equipment or diverting waste heat from electric generating plants can save energy, but such changes are costly. Many farmers may hesitate to make the financial investment necessary to save energy for fear of economic loss or unreliability. Even if a farmer decides to invest in energy-saving equipment, there may still be problems raising capital at an interest rate that makes investment economically justified. To overcome these barriers, states could offer loan programs to farmers who want to invest in energy-conserving but economically risky techniques.

A state could actively participate and make loans directly from the state treasury at favorable interest rates, or it could take a role that requires less financial commitment and administration by simply providing loan guarantees to private lenders. Each state must assess the costs and benefits of these loan programs before deciding to commit state funds to such projects.

Provide Tax Incentives for Investment in Energy-Saving Equipment. To encourage investment that will reduce energy consumption in greenhouse cultivation and frost protection, states can offer various tax incentives to farmers who purchase particular energy-saving equipment. Preferential tax treatment for investment in innovative energy-conserving devices can take several forms.

A farmer who invests in solar energy devices or in equipment that uses waste heat could be allowed to take an investment tax credit, or such equipment could be exempted from state or local taxes. Of course, a state can allow more than one type of tax incentive, depending on how much it wants to spur investment and how much tax revenue it is willing to forego.

CONCLUSION

Direct uses of energy on the farm account for about half of all the energy expended in agricultural production. Vast quantities of gasoline, diesel fuel, natural gas, and electricity are used to fuel the nations's farm machinery, irrigation pumps, crop dryers, greenhouses, and frost protection devices. Fortunately, farmers have many opportunities to increase their energy efficiency while maintaining farm productivity. For example, in their farm machinery operations farmers can reduce energy consumption by converting to diesel-powered machinery, improving maintenance and operating procedures, and minimizing tillage. Farmers who irrigate can improve energy efficiency by installing run-off reuse systems, scheduling irrigation, and maintaining irrigation pumps properly. Crop drying, particularly corn, is especially amenable to such energy-saving techniques as dryeration and solar drying. And greenhouse growers can lower fuel needs and operating expenses by engaging in such simple housekeeping improvements as adding insulation, adjusting thermostats properly, and sealing minor leaks.

States can actively promote the many energy-saving changes detailed in this chapter by implementing strategies geared to encourage energy conservation or to discourage energy waste. Whether these strategies include financial incentives, taxes, educational programs, demonstration projects, or regulations is, of course, a matter for each state to decide.

NOTES TO CHAPTER 3

1. U.S. Department of Agriculture, Economic Research Service, *The U.S. Food and Fiber Sector: Energy Use and Outlook*, prepared for the U.S. Senate Committee on Agriculture and Forestry (Washington, D.C.: U.S. Government Printing Office, 1974), pp. 6–7.

2. *Ibid.*, p. 5.

3. Clifford M. Hardin, "Foreword," in U.S. Department of Agriculture, *Contours of Change: Yearbook of Agriculture* (Washington, D.C.: U.S. Government Printing Office, 1970), p. xxxiii.

4. U.S. Department of Agriculture, Office of Communication, *Fact Book of U.S. Agriculture*, Miscellaneous Publication no. 1063 (Washington, D.C.: U.S. Government Printing Office, 1976), p. 2.

5. David Pimentel et al., "Food Production and the Energy Crisis," *Science* 182 (November 2, 1973): 444.

6. Wilson Clark, *Energy for Survival* (Garden City, New York: Anchor Press, 1974), p. 170.

7. Booz, Allen & Hamilton, Inc., *Energy Use in the Food System*, prepared for the Federal Energy Administration (Washington, D.C.: U.S. Government Printing Office, 1976), pp. IV−2, IV−19.

8. *Ibid.*, pp. IV−20, IV−23.

9. Using data based on energy consumption in 1963, Hirst concludes that agricultural production directly consumed 964 trillion BTUs, or 2 percent of total U.S. energy use. Eric Hirst, "Food-Related Energy Requirements," *Science* 184 (April 12, 1974): 136. The Steinharts estimate that over 1,300 trillion BTUs were directly expended in farm production in 1970, or 1.9 percent of total U.S. energy consumption. John S. Steinhart and Carol E. Steinhart, "Energy Use in the U.S. Food System," *Science* 184 (April 19, 1974): 309. And the U.S. Department of Agriculture's Economic Research Service calculates that the amount of energy directly consumed on farms in 1970 was 1,051 trillion BTUs, or about 1.6 percent of all U.S. energy used for that year. *The U.S. Food and Fiber Sector, supra* note 1 at xii–xv.

10. U.S. Department of Agriculture, *Energy to Keep Agriculture Going* (Washington, D.C.: U.S. Government Printing Office, 1974).

11. Dan Dvoskin and Earl O. Heady, *U.S. Agricultural Production Under Limited Energy Supplies, High Energy Prices, and Expanding Agricultural Exports* (Ames: Center for Agricultural and Rural Development, Iowa State University, 1976), pp. 100−101.

12. U.S. Department of Agriculture, Economic Research Service, *Changes in Farm Production and Efficiency*, Statistical Bulletin no. 561 (Washington, D.C.: U.S. Government Printing Office, 1976), p. 29.

13. *Fact Book of U.S. Agriculture, supra* note 4 at 3.

14. *Ibid.*, p. 13.

15. *Ibid.*, pp. 8−9; *The U.S. Food and Fiber Sector, supra* note 1 at 24.

16. Council for Agricultural Science and Technology, *Potential for Energy Conservation in Agricultural Production*, CAST Report no. 40 (Ames: Iowa State University, 1975), p. 5.

17. *The U.S. Food and Fiber Sector, supra* note 1 at 10.

18. Michael Perelman, "Farming with Petroleum, *Environment* 14, no. 8 (October 1972): 10.

19. CAST Report no. 40, *supra* note 16 at 5.

20. Many specific energy-conserving and cost-saving suggestions with regard to the use of farm machinery are contained in some recent booklets that are available from the USDA. Allen Schienbein, *A Guide to Energy Savings for the Field Crops Producer*, prepared for the U.S. Department of Agriculture and the Federal Energy Administration (Washington, D.C.: U.S. Department of Agriculture, 1977), pp. 38−49; N.A. Wynn, *A Guide to Energy Savings for the Vegetable Producer*, prepared for the U.S. Department of Agriculture and the Federal Energy Administration (Washington, D.C.: U.S. Department of Agriculture, 1977), pp. 24−35; N.A. Wynn, *A Guide to Energy Savings for the Orchard Grower*, prepared for the U.S. Department of Agriculture and the Federal Energy Administration (Washington, D.C.: U.S. Department of Agriculture, 1977), pp. 30−38.

21. Specific recommendations to modify tillage practices are described in Schienbein, *supra* note 20 at 10−14.

22. CAST Report no. 40, *supra* note 16 at 5.

23. U.S. Department of Agriculture, Economic Research Service, *1975 Changes in Farm Production and Efficiency*, Statistical Bulletin no. 548 (Washington, D.C.: U.S. Government Printing Office, 1975), p. 3.

24. *The U.S. Food and Fiber Sector*, *supra* note 1 at 30−31.

25. *Ibid.*, p. 26, *citing* National Market Reports, Inc., *National Farm Tractor and Implement Blue Book* (1974).

26. For example, between 1973 and 1975 the price of fifty- to fifty-nine-horsepower tractors rose from $6,480 to $8,790 and of ninety- to ninety-nine-horsepower tractors from $11,800 to $17,000. U.S. Department of Agriculture, Statistical Reporting Service, *Agricultural Prices Annual Summary 1975*, Publication Pr 1−3 (Washington, D.C.: U.S. Government Printing Office, June 1976), p. 123.

27. *See Fact Book of U.S. Agriculture*, *supra* note 4 at 13.

28. *Changes in Farm Production and Efficiency*, *supra* note 12 at 2.

29. CAST Report no. 40, *supra* note 16 at 5.

30. *Ibid.*, p. 6.

31. *Ibid.*

32. *Ibid.*, p. 7.

33. 15 U.S.C. § 2009 (Supp. V 1976).

34. University of Nebraska, Department of Agricultural Engineering, *Nebraska Tractor Test Data* (Lincoln: University of Nebraska, College of Agriculture, 1974).

35. CAST Report no. 40, *supra* note 16 at 5.

36. *Ibid.*

37. "How Plowless Farming Saves the Soil," *Business Week*, August 16, 1976, p. 110.

38. "Farming Without the Plow," *Washington Post*, January 18, 1976, p. A−1. For a general discussion of farming without tillage, see Glover B. Triplett, Jr., and David M. Van Doren, Jr., "Agriculture without Tillage," *Scientific American* 236, no. 1 (January 1977): 28−33.

39. "Farming Without the Plow," *supra* note 38 at A−1.

40. CAST Report no. 40, *supra* note 16 at 7.

41. *See* Earle Gavett, "Agriculture: Energy Use and Conservation" (Speech presented at Texas A&M University, College Station, Texas, 1973), p. 28.

42. *Ibid.*, p. 13.

43. C.G. Coble and W.A. LePori, *Energy Consumption, Conservation and Projected Needs for Texas Agriculture*, Publication S/D−12, Special Project B, prepared for the Governor's Energy Advisory Council (College Station: Texas Agricultural Experiment Station, 1974), p. 38.

44. D.R. Griffith, J.V. Mannering, and C.B. Richey, "Energy Requirements and Areas of Adaptation for Eight Tillage-Planting Systems for Corn," in *Agriculture and Energy*, ed. William Lockeretz (New York: Academic Press, 1977), p. 274.

45. CAST Report no. 40, *supra* note 16 at 8; "How Plowless Farming Saves the Soil," *supra* note 37 at 110.

46. Lehi German et al., "Economic and Energy Efficiency Comparisons of Soybean Tillage Systems," in *Agriculture and Energy*, ed. William Lockeretz (New York: Academic Press, 1977), p. 285.

47. See "How Plowless Farming Saves the Soil," *supra* note 37 at 113.

48. CAST Report no. 40, *supra* note 16 at 8.

49. See "Farming Without the Plow," *supra* note 38 at A-4.

50. CAST Report no. 40, *supra* note 16 at 7.

51. Lester R. Brown and Erik P. Eckholm, *By Bread Alone* (New York: Praeger, 1974), p. 92.

52. Dan Dvoskin, Ken Nicol, and Earl O. Heady, "Irrigation Energy Requirements in the 17 Western States," in *Agriculture and Energy*, ed. William Lockeretz (New York: Academic Press, 1977), p. 103. The fifty-four million irrigated acres represented 15 percent of the 367 million acres of cropland used for crops in 1975. *Changes in Farm Production and Efficiency*, *supra* note 12 at 17.

53. See Toups Corporation, *Water Pollution Abatement Technology: Capabilities and Costs: Irrigated Agriculture*, prepared for the National Commission on Water Quality (Springfield, Virginia: U.S. Department of Commerce, National Technical Information Service, 1976), p. 2.

54. Dvoskin, Nicol, and Heady, *supra* note 52 at 104. The seventeen western states, in order of irrigated acreage, are California, Texas, Nebraska, Colorado, Idaho, Montana, Wyoming, Kansas, Oregon, Washington, Arizona, Utah, New Mexico, Nevada, Oklahoma, South Dakota, and North Dakota.

55. Edward Groth III, "Increasing the Harvest," *Environment* 17, no. 1 (January-February 1975): 29; Toups Corp., *supra* note 53 at 9.

56. Toups Corp., *supra* note 53 at 9.

57. Dvoskin, Nicol, and Heady, *supra* note 52 at 106; "Agriculture: Irrigation Costs, Availability Seen Bearing Heavily on Outlook for Farming," *BNA Energy Users Rep.* no. 107 (August 28, 1975): A-4, A-5.

58. P.E. Fischbach, J.J. Sulek, and D.D. Axthelm, *Your Pumping Plant May Be Using Too Much Fuel* (Lincoln: University of Nebraska, College of Agriculture and Home Economics Extension Service, 1973).

59. Dvoskin, Nicol, and Heady, *supra* note 52 at 107.

60. *Ibid.*, p. 108.

61. William E. Splinter, "Center-Pivot Irrigation," *Scientific American* 234, no. 6 (June 1976): 90-99.

62. *Ibid.*, p. 94.

63. Dvoskin, Nicol, and Heady, *supra* note 52 at 109.

64. Gordon Sloggett, "Energy Used for Pumping Irrigation Water in the United States, 1974," in *Agriculture and Energy*, ed. William Lockeretz (New York: Academic Press, 1977), p. 113.

65. Steinhart and Steinhart, *supra* note 9 at 309.

66. James R. Gilley and Darrell G. Watts, "Energy Reduction Through Improved Irrigation Practices," in *Agriculture and Energy*, ed. William Lockeretz (New York: Academic Press, 1977), p. 188.

67. Splinter, *supra* note 61 at 94.

68. D.E. Lane, P.E. Fischbach, and N.C. Teter, *Energy Uses in Nebraska Agriculture* (Lincoln: University of Nebraska, College of Agriculture and Home Economics Extension Service, 1973), p. 17.

69. The U.S. Department of Agriculture and the Federal Energy Administration have compiled a set of six excellent booklets that describe energy-saving practices. Three of these publications contain suggested techniques and potential cost savings for farmers who irrigate. *See* Schienbein, *supra* note 20 at 19–27; Wynn, *A Guide to Energy Savings for the Vegetable Producer, supra* note 20 at 18–23; and Wynn, *A Guide to Energy Savings for the Orchard Grower, supra* note 20 at 11–17.

70. Gilley and Watts, *supra* note 66 at 189–90.

71. *Ibid.*, p. 190.

72. *Ibid.*, p. 193. *See also* Arland D. Schneider and Leon New, "Energy Requirements Low for Tailwater and Lake Pumps," *The Cross Section* 21, no. 2 (February 1975): 1, 3, 4.

73. Gilley and Watts, *supra* note 66 at 190.

74. *Ibid.*

75. *Ibid.* For a general discussion of trickle irrigation, *see* Kevin P. Shea, "Irrigation Without Waste," *Environment* 17, no. 5 (July–August 1975): 12–15.

76. CAST Report no. 40, *supra* note 16 at 19.

77. Gilley and Watts, *supra* note 66 at 192.

78. "Demand Control Systems Save Money for Rural Utilities and Their Customers," *BNA Energy Users Rep.* no. 211 (August 25, 1977): 13–14.

79. Gilley and Watts, *supra* note 66 at 195.

80. Fischbach, Sulek, and Axthelm, *supra* note 58. *See also* Gilley and Watts, *supra* note 66 at 192, which describes the results of another study where the irrigation pumps were operating even less efficiently, at 25 percent below the Nebraska Performance Standard.

81. Coble and LePori, *supra* note 43 at 32–35.

82. CAST Report no. 40, *supra* note 16 at 19.

83. Gilley and Watts, *supra* note 66 at 200.

84. *See* National Association of State Universities and Land-Grant Colleges and U.S. Department of Agriculture, *A National Program of Agricultural Energy Research and Development*, prepared for the National Planning Committee of the Agricultural Policy Advisory Committee (Washington, D.C.: U.S. Department of Agriculture, 1976); Food and Agriculture Act of 1977, 7 U.S.C.A. § 3101(8)(E) (West Supp. February 1978).

85. The funding consists of $300,000 from ERDA, $100,000 from the New Mexico Energy Resources Board, $50,000 from the Interstate Stream Commission, and $50,000 from the Four Corners Commission. *See Solar Energy Digest* 7, no. 2 (August 1976): 2.

86. Nebraska Ground Water Management Act, Neb. Rev. Stat. §§ 46–656 to 46–674 (Cum. Supp. 1976).

87. *Id.* at § 46–664.

88. Special Comm. on Energy and Natural Resources, "Re: Proposal No. 14—Groundwater Use," *Reports of Special Committees of the 1976 Kansas Legislature*, p. 226.

89. Ch. 114, 1977 N.M. Laws (to be codified in N.M. Stat. Ann. § 72–15A–11.4). The complete text of the act is set out below.

Section 1. A new Section 72–15A–11.4 NMSA 1953 is enacted to read: "72–15A–11.4. CREDIT AGAINST PERSONAL INCOME TAX—REFUND.—

A. Any resident who files an individual New Mexico income tax return and who is not a dependent of another taxpayer may claim a tax credit in an amount not to exceed twenty-five thousand dollars ($25,000) of the cost of equipment used for construction of a solar energy system for irrigation pumping purposes and used on the taxpayer's real property located in New Mexico. The person furnishing the equipment shall furnish the taxpayer with an accounting of the cost of the equipment to the taxpayer.

B. The taxpayer may claim the credit if:

(1) he submitted a plan and design of the solar energy system for irrigation pumping purposes to the energy resources board prior to installation and received approval from the energy resources board of the plan or design; and

(2) received certification from the energy resources board after installation of the system that it conformed to the design or plan submitted and that it will result in at least a seventy-five percent reduction in the utilization of energy produced by fossil fuels or in the utilization of secondary forms of energy dependent upon fossil fuels for its generation.

C. A taxpayer may claim the credit provided by the provisions of this section for each taxable year in which solar energy equipment is installed. Claims for the credit provided in this section shall be limited to three consecutive years and the maximum aggregate credit allowable shall not exceed twenty-five thousand dollars ($25,000) for any single solar energy project.

D. A taxpayer may not claim the credit provided by the provisions of this section if he has claimed a similar credit, deduction, exemption or exclusion on his federal income tax return under a provision that relates to the construction of a non-fossil fuel or solar energy system or if he has claimed a credit under Section 72–15A–11.3 NMSA 1953 for such equipment.

E. A husband and wife who file separate returns for a taxable year in which they could have filed a joint return may each claim only one-half of the tax credit that would have been allowed on a joint return.

F. A taxpayer who otherwise qualifies and claims a credit on a project that was constructed by a partnership of which the taxpayer is a member may claim a credit only in proportion with his interest in the partnership. The total credit claimed by all members of the partnership shall not exceed twenty-five thousand dollars ($25,000) in the aggregate.

G. The credit provided by this section may only be deducted from the taxpayer's New Mexico income tax liability for the taxable year in which the equipment was installed on the taxpayer's property. If the tax credit exceeds the taxpayer's income tax liability, the excess shall be refunded to the taxpayer."

Section 2. APPLICABILITY.—The provisions of this act apply to taxable years beginning on or after January 1, 1977.

90. Specifically exempted from the transaction privilege tax and the use tax are solar energy devices, defined as "A system or series of devices designed primarily to provide heating or cooling or both, or to produce electrical or mechanical power or both, or to pump irrigation water by means of collecting and transferring solar generated energy including devices having the capacity for storing solar energy." Ariz. Rev. Stat. Ann. §§ 42–1312.01(A)(9), 42–1409 (B)(9) (Supp. 1977).

91. Earl O. Heady et al., *Agricultural and Water Policies and the Environment: An Analysis of National Alternatives in Natural Resource Use, Food Supply Capacity, and Environmental Quality*, CARD Report no. 40T (Ames: Center for Agricultural and Rural Development, Iowa State University, 1972).

92. National Water Commission, *Water Policies for the Future* (Washington, D.C.: U.S. Government Printing Office, 1973).

93. Dvoskin and Heady, *supra* note 11 at 15.

94. *Ibid.*, p. 101.

95. The State of Washington was stymied in its attempt to impose a charge for irrigation water. *See Farm Journal* 100, no. 3 (Mid-February 1976): 23–24.

96. Vashek Cervinka et al., *Energy Requirements for Agriculture in California*, prepared for the California Department of Food and Agriculture and the University of California, Davis (n.p., 1974), p. 16.

97. Milvern H. Noffke, Deon D. Axthelm, and H.R. Mulliner, *The Benedict Project* (York, Nebraska: Blue River Association of Groundwater Conservation Districts, 1975).

98. *Ibid.*, p. 3.

99. "Demand Control Systems Save Money for Rural Utilities and Their Customers," *supra* note 78 at 13–14.

100. *Ibid.*

101. *Ibid.*

102. Indeed, for safe, long-term storage and for some overseas shipment, the moisture content must be 13.5 to 14 percent or less. *See* Otto C. Doering III and Bruce A. McKenzie, *Agriculture and the Energy Dilemma*, CES Paper no. 5 (West Lafayette, Indiana: Purdue University, Cooperative Extension Service, 1973), p. 3.

103. *Ibid.*

104. *Ibid.*

105. John D. Buffington and Jerrold H. Zar, "Realistic and Unrealistic Energy Conservation Potential in Agriculture," in *Agriculture and Energy*, ed. William Lockeretz (New York: Academic Press, 1977), p. 699.

106. CAST Report no. 40, *supra* note 16 at 10. A 1974 study estimated that the amount of energy used to dry corn increased from the equivalent of 96,800 BTUs an acre in 1945 to almost 1.6 million BTUs an acre in 1959 to over 2.9 million BTUs an acre in 1970. *See* David Pimentel et al., *Workshop on Research Methodologies for Studies of Energy, Food, Man and Environment, Phase I* (Ithaca: Center for Environmental Quality Management, Cornell University, 1974), p. 20, *revising data originally published in* David Pimentel et al., "Food Production and the Energy Crisis," *Science* 182 (November 2, 1973): 443–49.

107. *The U.S. Food and Fiber Sector, supra* note 1 at 14.

108. *Ibid.*

109. *See* Schienbein, *supra* note 20 at 28–34.

110. Doering and McKenzie, *supra* note 102 at 5.

111. Council for Agricultural Science and Technology, *Energy in Agriculture*, CAST Report no. 14 (Ames: Iowa State University, 1973), p. 10.

112. CAST Report no. 40, *supra* note 16 at 11–12, discusses various drying methods and concludes that a system combining high temperature drying with dryeration has the greatest drying efficiency.

113. *Ibid.*

114. U.S. Energy Research and Development Administration, *Solar Energy for Agriculture and Industrial Process Heat*, ERDA Publication no. 76–88 (Washington, D.C.: U.S. Government Printing Office, 1976), p. 3.

115. George H. Foster and Robert M. Peart, *Solar Grain Drying: Progress and Potential*, Agricultural Information Bulletin no. 401, prepared for the Agricultural Research Service, U.S. Department of Agriculture (Washington, D.C.: U.S. Government Printing Office, 1976).

116. Utilizing a solar dryer to provide supplemental heat, the author of the study found that in the first year, even with a late start, solar energy contributed 50 percent of the total energy needed to heat the air and almost 32 percent of the total energy required to dry the corn. When 3,800 bushels of corn were dried in this way, the yearly saving was $146 on an investment of $465. *See* Barry S. Bauman, Marshall F. Finner, and Gene C. Shove, *Low Temperature Grain Drying with Supplemental Solar Heat from an Adjacent Metal Building*, Paper no. 75–3514 (St. Joseph, Michigan: American Society of Agricultural Engineers, 1975), pp. 6–7.

117. "Solar Grain Drying . . . Built into Your System," *Farm Journal* 100, no. 8 (August 1976): 20, 22. For the results of three farmers who adopted solar grain drying, *see* "The Sun: At Home on the Farm," *Solar Age* 3, no. 2 (February 1978): 28–31.

118. For example, scientists and engineers in eleven states are researching potential uses of solar energy in grain drying under a $300,000 grant from ERDA. The research is being coordinated by two agencies within USDA, the Agricultural Research Service and the Cooperative State Research Service. *See BNA Energy Users Rep.* no. 127 (January 15, 1976): D–3, D–4; *Solar Energy for Agriculture and Industrial Process Heat, supra* note 114 at 3–4.

119. Cattle show a 5 to 7 percent increase in weight gain when fed wet shelled corn rather than dry meal and a 10 to 12 percent increase in weight gain on a diet of wet corn and cob mixture rather than dry meal. CAST Report no. 40, *supra* note 16 at 13.

120. Hogs exhibit a 3 to 5 percent decrease in weight gain when fed wet corn rather than dry meal. Moreover, feeding wet corn to hogs is limited to animals that weigh more than seventy pounds. *Ibid.*

121. Evidence that Congress wants the Cooperative Extension Service to provide practical information on the uses of solar energy in agricultural production can be found in the Food and Agriculture Act of 1977, which amended the law regarding cooperative agricultural extension work. 7 U.S.C.A. §§ 341, 342 (West Supp. February 1978).

122. The Cooperative Extension Service at Purdue University, a land-grant university in an agricultural state, actively provides energy conservation advice to farmers in the course of consultations and farm management workshops. Purdue's Department of Agricultural Engineering Extension has conducted energy conservation workshops on corn dryer operation. For information, the person to contact is:

> Dr. Otto C. Doering III
> Department of Agricultural Economics
> Purdue University
> Lafayette, IN 47907
> (317) 493-2180

123. The costs run from 50 cents to $5 a square foot. "Solar Grain Drying . . . Built into Your System," *supra* note 117 at 22.

124. CAST Report no. 40, *supra* note 16 at 22.

125. Cervinka et al., *supra* note 96 at 148-49.

126. *Ibid.*

127. *See* Wynn, *A Guide to Energy Savings for the Vegetable Producer*, *supra* note 20 at 36-45.

128. This list of measures is derived from CAST Report no. 40, *supra* note 16 at 22, and from G.B. Taylor, "Agricultural Energy Use," in *Efficient Electricity Use*, ed. Craig B. Smith, prepared for the Electric Power Research Institute (New York: Pergamon Press, 1976), p. 402.

129. CAST Report no. 40, *supra* note 16 at 22.

130. James G. Knudsen and Larry L. Boersma, eds., *Future Developments in Waste Heat Utilization*, Circular 49 (Corvallis: Oregon State University, Engineering Experiment Station, 1975), pp. 63-85.

131. *Solar Energy for Agriculture and Industrial Process Heat, supra* note 114 at 3.

132. *See, e.g.*, James C. McCullagh, ed., *The Solar Greenhouse Book* (Emmaus, Pennsylvania: Rodale Press, 1978).

133. *See* Wynn, *A Guide to Energy Savings for the Orchard Grower*, *supra* note 20 at 25-29.

134. Cervinka et al., *supra* note 96 at 135-38.

135. CAST report no. 40, *supra* note 16 at 19-20. Perhaps surprisingly, CAST reports that gasoline-powered wind machines use approximately four times as much energy an acre each year as do electrically operated wind machines.

136. To promote our national agricultural and energy policies, Congress has directed the secretary of agriculture to conduct research and development relating to uses of solar energy in farm buildings and farm machinery. *See* 7 U.S.C.A. § 427 (West Supp. February 1978). For specific examples of agricultural research regarding energy conservation in greenhouses, *see Solar Energy Intelligence Report* 2, no. 17 (August 16, 1976): 136.

Chapter 4

Reducing Indirect Uses
of Energy on Farms

In the early days of our nation's agriculture, land was abundant and labor cheap. To till the land and harvest the crops, farmers combined the labor of man and beast, using farm machinery only sparingly. Soil fertility was maintained through use of animal wastes and crop rotation. Today, most farmers substitute tractors for draft animals, fossil fuels for animal feeds, and to a large extent, chemical fertilizers for manure and nitrogen-fixing crops. The plain fact is that American farmers have traded hours of labor for barrels of oil to produce bushels of grain.

The gradual changeover from labor-intensive to capital-and energy-intensive agriculture in the United States has resulted in the expansion of an entire agricultural industry, often called agribusiness. One part of the agribusiness complex is the large-scale modern farm where much of our nation's food and fiber is produced. These farms, in turn, are fed by the energy-intensive manufacturing industries that supply the farmers with seed, machinery, refined petroleum products, fertilizers, and pesticides.

When farmers began to substitute these manufactured goods for items once produced on farms, a number of major changes were set in motion. As capital was substituted for labor, farmers registered the impressive gains in food and fiber production that have made American agriculture the most productive in the world in both total output and output per man-hour. The replacement of labor with capital proved so economically efficient that many farmers were displaced and left the farm for the city. Today, workers who produce the agricultural inputs outnumber farmers by about two to one.

103

This increased dependence on agricultural inputs manufactured away from the farm, however, has resulted in higher production costs, as the prices of tractors, fuels, fertilizers, and pesticides have increased dramatically in the last few years. These higher operating costs have made farmers more vulnerable to fluctuations in the prices paid for inputs and the prices received for commodities. Relying on agricultural inputs is, in effect, relying on nonrenewable fossil fuel energy. Some commentators have noted that although total crop production has increased, the energy efficiency of crop production, measured in terms of energy output per unit of energy input, has declined.[1] Taken together, the increase in total energy use and the decrease in energy efficiency mean that when energy prices go up or energy supplies go down, the farmer feels the pinch.

In the previous chapter, the discussion focused on the farming operations that consume vast quantities of petroleum products and electricity. The energy used to plant, cultivate, irrigate, harvest, and dry crops is termed direct consumption. Farmers use even more energy in the form of chemical fertilizers, synthetic pesticides, hybrid seeds, and high protein feed supplements. The energy used in the manufacture of these inputs is termed indirect consumption. Additional energy is expended in the production of farm machinery, agricultural steel, and refined petroleum products, but these items are usually included in a separate category called capital inputs.[2]

Those who have studied the food production system conclude that the amount of energy expended indirectly is slightly greater than what is expended directly. In terms of absolute amounts, this means that farmers use at least 750 trillion and probably closer to 1,500 trillion BTUs for fertilizers, pesticides, seed, feed, and the like. In terms of national energy expenditures, this amount translates into approximately 1 to 2 percent of all energy used in the nation.[3]

This chapter focuses on the energy used to manufacture two important inputs, fertilizers and pesticides. The discussion will concentrate on strategies that individual states can adopt to increase the energy efficiency of fertilizer and pesticide application. Focusing on energy conservation in the use of these two inputs is appropriate for several reasons. The large amount of energy used in fertilizer production makes it a logical target for energy-conserving strategies. Manufacturing fertilizers, chiefly nitrogen fertilizers, uses the lion's share of energy that is embodied in input materials—over 60 percent of all the energy used to produce inputs.[4] Indeed, the energy embodied in the fertilizers applied on farms in the United States today is second only to the energy contained in the fuels that power all farm machinery.[5] In addition, farmers have several alternatives from which to choose in order to conserve energy in fertilizing their crops. This ease of change is another factor that favors efforts to conserve energy in fertilizer application.

The second section of this chapter discusses how energy can be conserved in controlling pests. Although the production of pesticides uses much less energy than the manufacture of fertilizers, the unnecessary use of pesticides poses a greater potential hazard to the environment. The importance of reducing unnecessary pesticide application, then, is primarily to limit the risk of harm to people and the environment and, second, to cut back on energy consumption.

FERTILIZERS

Most agricultural researchers will agree on two main reasons for the large increases in U.S. crop yields over the past three decades: the development of hybrid plant varieties and the increased use of chemical fertilizers. This section focuses on fertilizers, in particular the reasons for fertilizer use, the energy costs of producing and using manufactured fertilizers, some suggested energy-conserving changes, and some strategies to implement those changes.

Introduction

Crops cannot grow healthy and vigorous without adequate amounts of the nutrients they need. Of the eighteen nutrients indispensable to plant health, three primary nutrients—nitrogen (N), phosphorus (P), and potassium (K)—account for almost all the active components of fertilizers used in U.S. crop production.

If crops were simply left in the fields to decompose, the nutrients that had been extracted from the soil in growing would be returned to the soil, where the level of nutrients would remain in balance. Since, however, the goal of agriculture is to harvest what is grown, nutrients taken from the soil must be continually replenished to maintain the fertility of the land. How much and how often to fertilize depend on the nutritional needs of the crop and on the natural level of nutrients in the soil. Corn, for example, requires a large quantity of nutrients to produce high yields. A corn yield of 135 bushels an acre removes from the soil an average of 150 pounds of nitrogen, 75 pounds of phosphorus, and 75 pounds of potassium.[6] Leguminous crops like soybeans and alfalfa produce much of the nitrogen they need and so can be grown with only modest additions of nitrogen fertilizer.

In the 1940s, drastic changes were set in motion when a series of price decreases led farmers to step up their use of commercially prepared fertilizers. Use of store-bought fertilizers increased 129 percent between 1940 and 1950, 69 percent between 1950 and 1960, and 113 percent between 1960 and 1970.[7] Average fertilizer application for a corn-producing farm zoomed from 19 pounds an acre in 1945 to 203 pounds an acre in 1970.[8] Total fertilizer use for all farms more than quadrupled from 4 million tons in 1950 to a peak of 19.3 million tons in 1974 before declining to 17.6 million tons in 1975.[9]

In the twentieth century, the substitution of low cost energy for land and labor has unleashed the full potential of food production. The results of this heavy use of commercially prepared fertilizers, along with other scientific and technological advances, can only be called remarkable. Average corn yields rose from thirty-four bushels an acre in 1945 to a high of ninety-seven bushels an acre in 1972. Since then, yields have declined somewhat, but nevertheless, corn production has remained spectacular. Farmers in 1975 produced more than double the 2.7 billion bushels harvested in 1950.[10] As production increased, total acreage under cultivation remained fairly constant until 1973, when additional cropland was brought into use. Although fertilizer use alone cannot be credited with the dramatic gains in farm productivity—improvements in hybrid crop varieties and increased use of farm machinery, irrigation, and pesticides all contributed—agricultural experts estimate that fertilizer use accounts for fully one-third to one-half of current agricultural production in the United States.[11]

The Fertilizer Institute estimates that more than 505 trillion BTUs went into the production of the 19.3 million tons of fertilizer applied to crops in 1974.[12] Of the three major nutrients in chemical fertilizer, nitrogen has the largest energy budget. Producing a pound of nitrogen consumes approximately 25,000 BTUs. Thus, the 9.1 million tons of nitrogen produced in 1974 represent an energy expenditure of 455 trillion BTUs.[13] The amounts of energy needed to manufacture phosphorus and potassium, though much smaller than for nitrogen, are still enormous. Into the production of a pound of phosphorus go approximately 3,000 BTUs, while the manufacture of potassium consumes about 2,000 BTUs a pound. In 1974, 5.1 million tons each of phosphorus and potassium were processed, representing an energy expenditure of 51 trillion BTUs.[14]

The main source of energy for manufacturing chemical fertilizers is natural gas. Natural gas supplies an estimated 70 to 80 percent of all the energy that goes into fertilizer production.[15] Playing a dual role, natural gas serves both as a source of hydrogen in synthesizing anhydrous ammonia (a key nitrogen fertilizer) and as a fuel to produce process heat in manufacturing all fertilizers. No substitute for natural gas as a feedstock in nitrogen fertilizer production is satisfactory to maintain current production levels and current prices without major changes in plant design. Although natural gas could be replaced as a source of process heat by oil or coal, natural gas is the preferred choice because it is currently the least expensive.

When energy costs were low, the use of energy-intensive chemical fertilizers made good economic sense. Now that energy prices are rising rapidly with no limit in sight and our energy supplies are increasingly being met by imports, alternatives to the use of energy-intensive fertilizers must be considered.

The first part of the problem is based on economics. Fertilizer prices skyrocketed in 1974 and climbed still higher in 1975. Since mid-1975, fertilizer prices have declined somewhat back to 1974 levels, but are still more than double their price in autumn 1973.[16] A study prepared for the Federal Energy Administration lays blame for the fertilizer price increase partly on the elimination of price controls and partly on the imbalance between fertilizer supply and demand.[17] Regardless of why prices rose so high, it is unlikely that fertilizer prices will ever again be as low as they were prior to the Arab oil embargo and the onset of the "energy crisis." Increased costs already have forced farmers to cut back on the amounts of fertilizer they buy and apply. In 1975, for the first time in nearly twenty years, the use of commercially prepared fertilizer declined.[18]

Like all else, fertilizer effectiveness must bow to the law of diminishing returns. As more and more fertilizer is applied to a crop, the yield continues to increase but at a decreasing rate. For example, the first forty pounds of nitrogen applied to a field of corn produces twenty-seven additional pounds of corn for each pound of nitrogen; another forty pounds of nitrogen produces only fourteen additional pounds of corn for each pound of nitrogen added; and so on as the average increase in corn yield produced by successive pounds of nitrogen diminishes.[19] As the prices of fertilizers, especially nitrogen fertilizer, increased, farmers cut back fertilized acreage and decreased the rate of fertilizer application purely for economic reasons. The marginal costs of applying additional fertilizer simply exceeded the marginal revenues of increased crop yield.

Environmental pollution is a second problem calling for reduced use of chemical fertilizers. Fertilizer manufacturing plants pollute the air and water around them, often severely.[20] Moreover, fertilizers are a source of nitrate pollution of groundwater, a problem that is compounded by farmers who apply nitrogen fertilizer at a rate in excess of what the plants can absorb. In general, crops are seldom able to use more than 50 to 60 percent of nitrogen applied to the soil, and the rest is lost to the atmosphere or to run-off waters.[21]

Finally, some people maintain that chemical fertilizers, as opposed to natural fertilizers, may have adverse effects on soil quality. Chemical fertilizers do not replenish the humus content of the soil, which may cause it to become less and less permeable to air and water. The result is excess water run-off and soil erosion. Under such conditions, the efficiency of plant nutrient uptake is bound to deteriorate.[22]

High crop yields depend on fertilization, yet continued reliance on chemical fertilizers is acceptable neither economically nor ecologically. Greater capital and labor inputs, less energy-intensive fertilizers, and alternative farming methods all can help take up where chemical fertilizers leave off, enabling farmers to keep up their profits as they cut down their total energy consumption.

Suggested Changes and Strategies

Farmers can conserve energy in their application of fertilizers by adopting various practices, some old and some new.[23] A recent study prepared for the Electric Power Research Institute presents some options that can serve as an introduction to this discussion of suggested changes and strategies for change. Several energy-conserving steps and estimates of the percentage savings possible are outlined in Table 4—1, which is reprinted from that study.

Table 4—1.   Energy-Saving Tips for Fertilizer Use

• Have a soil analysis run each year	10–20%
• Conserve the bulk of fertilizer supplies for application to lower-testing soils, or to prorate over a greater acreage and on crops which will make the maximum use of fertilizer	10–15%
• Starter fertilizer applied through the planter or drill is more efficient than broadcast fertilizer	10–12%
• Adjust nitrogen application to the date of planting	8–10%
• Include credit for manure in rates adjustment	5–10%
• Question whether fall application of nitrogen is right for the soil, and the likely losses from leaching	5– 8%

Estimated percentage savings based on total expenditure of energy use. Actual savings will vary depending on specific applications, but should fall in these ranges. These savings are not necessarily additive.

Source: G.B. Taylor, "Agricultural Energy Use," in *Efficient Electricity Use*, ed. Craig B. Smith, prepared for the Electric Power Research Institute (New York: Pergamon Press, 1976), p. 404.

Match Fertilizer Use to Soil Conditions and Plant Needs. In accordance with basic economic theory, farmers will try to use just enough fertilizer to maximize profits. When fertilizer is applied at the most economically efficient level, the marginal cost of adding one more unit of fertilizer equals the value of the additional yield resulting from the application of that unit of fertilizer. Using any less fertilizer would mean less than optimal crop yields in relation to fertilizer cost and thus profits lost; using any more fertilizer would mean only modestly increased crop yields insufficient to cover the additional fertilizer costs and, again, profits lost. Both the farmer and society have an interest in preventing overfertilization of the land.

In determining the optimum level of fertilizer application for a particular crop, two considerations are crucial: the needs of the crop and the condition of the soil in the area. Of course, different crops require different amounts of nutrients. Corn, for example, requires heavy application of nitrogen supplements, whereas soybeans and other legumes produce much of the nitrogen they need.

Using soil tests, farmers can monitor the balance of nutrients in the soil and so determine what fertilizers to apply for a particular crop and how much to apply.[24] Properly performed and interpreted, soil tests are most effective guides to the economically efficient use of fertilizers. When fertilizers are scarce, soil tests can help farmers choose how to distribute the limited supply of nutrients for the greatest economic return. Soil testing procedures cannot, however, accurately predict how much of the nitrogen present in a particular soil is in a form useful to crops.

Some researchers who have studied the past results of soil testing programs suggest that the widespread adoption of soil testing is unlikely to result in significant energy savings, since many farmers presently apply too little rather than too much fertilizer.[25] Their conclusion seems overly pessimistic. Surely farmers who discover that they are overfertilizing will eliminate unnecessary use of this expensive and energy-intensive product. Soil tests will encourage them to match fertilizer use more closely to the needs of their crops, with a resulting decrease in total energy consumption. Farmers who find that they are underfertilizing their crops may decide to increase their application of fertilizer to attain optimal yields, but with fertilizers that are less energy-intensive than commercially prepared fertilizers. These farmers can be encouraged to save energy by using natural methods of fertilization and by adopting techniques that help crops absorb as much as possible the fertilizers that are applied.

To promote the use of soil testing, experts from state departments of agriculture, land-grant colleges, or industry could properly test soil and interpret test results for farmers. State-supported and commercial soil-testing laboratories are now operating in every state. Despite the availability of testing facilities, the use of these facilities has been disappointing: only 2.1 million soil samples were tested in 1973, an average of one sample for each 162 acres of cropland planted.[26]

To expand the use of soil tests, state governments could either do the tests directly or pay outside experts to do them. Fertilizer companies have offered free soil tests in the past and may continue the practice. But many farmers are skeptical about industry testing, suspecting ulterior motives. Preferably, then, government or university experts should do the testing.

Educational programs, such as those offered by agricultural extension services, offer another means of getting farmers to use fertilizers economically. Programs could emphasize the importance and potential benefits of soil tests, working in combination with a program of government-subsidized soil tests. The small cost of providing these services would likely be more than offset by gains in production, in profits to farmers, and in energy savings.

Improve the Timing and Method of Fertilizer Application. Once the fertilizer needs have been matched to soil and crop characteristics, farmers must decide when and how to apply it. When application is properly timed, nutrients are provided to plants as needed with minimal nutrient loss or inactivation. Timing is crucial, particularly for expensive, energy-intensive nitrogen fertilizers easily lost into the drainage water by leaching or into the atmosphere by volatilization. The method of application also can influence how well crops make use of fertilizer. By minimizing fertilizer losses through controlled timing and method of application, farmers can save energy.

Because the cost of nitrogen fertilizer was relatively low until recent years, some farmers became careless about timing their fertilizer applications. Putting on fertilizer in the fall frees up some time in the spring, but energy-efficiency and fertilizer effectiveness are sacrificed for convenience and time. Now that fertilizer costs are high and energy conservation is a national concern, the time has come for reevaluation. For nitrogen especially, the time of application is critical in terms of both plant uptake and nutrient loss. Although plants can best utilize phosphorus applied at or near the time of planting, virtually none is likely to be lost through leaching or volatilization. Timing of potassium application can vary with little effect on the efficiency of plant uptake or the degree of nutrient loss. The Council for Agricultural Science and Technology (CAST) estimates that nitrogen application to corn alone could be reduced 5 percent by better timing, representing an annual saving of approximately 192,000 tons, or about 10 trillion BTUs a year.[27]

Applying nitrogen fertilizer in the spring, just before planting, rather than in both autumn and spring, would save fertilizer and reduce energy consumption. This simple suggestion is, however, laden with potential problems. Farmers must be assured of ample supplies of nitrogen available at or near planting time; such a reliance on fertilizer suppliers can be risky and costly. Instead, farmers could build storage facilities for nitrogen fertilizer, although the capital investment would be substantial. Applying fertilizer in the spring only also creates a time squeeze. So many essential tasks must be accomplished in such a short time that any delay can mean decreased crop yields and reduced profits. Moreover, since potential nitrogen loss depends on soil conditions and climate, spring application of nitrogen fertilizer may not be workable in every part of the country. Where soil is coarse and heavy spring rains come before crops have developed good roots, nitrogen losses can be great. The problem of leaching is most serious in the Southeast, the Northeast, and the southern Corn Belt, and it is in these areas of the country that the potential is greatest for fertilizer savings through improved timing of fertilizer applications. This method will be less useful in the central and northern Corn Belt and other areas of the country where soils remain frozen from late fall until early spring, thereby limiting nitrogen loss.

The method of application also affects fertilizer use and energy consumption. Machinery used to apply fertilizer should be carefully calibrated, since applying too much fertilizer is not only a waste but also may do more harm than good. Fertilizer should be applied as close as possible to the plant root zone without damaging the plant. Farmers may find that band application results in better yields than broadcast application. In general, fertilization should be combined with some other operation whenever possible. Farmers who irrigate can often apply supplemental nitrogen through the irrigation water to reduce energy consumption and production costs.

Recently developed nitrification inhibitors slow down the conversion of ammonium (a usable form of nitrogen fertilizer) to nitrate (an unusable form). Reducing the rate of nitrification can mean that nitrogen applied to crops remains available to them longer. Nitrification inhibitors have also been found to augment the protein content of crops and to prevent some crop disease.[28] Recent experiments suggest that adding nitrification inhibitors to all the nitrogen fertilizer applied to wheat and corn in the United States could save up to 22 trillion BTUs by cutting nitrogen losses up to 10 percent.[29] Nitrification inhibitors make possible fall application of fertilizers, eliminating the spring application, a step that could save over 2 trillion BTUs.[30] Because crops recover a greater proportion of the nitrogen applied, crop yields are higher, promising an additional energy savings of over 4 trillion BTUs.[31] Nitrification inhibitors thus promise increasingly efficient nitrogen use for the future.

Ideas for achieving better timing and methods of fertilizer application could be disseminated widely through information transfer programs. Since many elements—soil, climate, geography, crop grown, fertilizer type, and method of application—enter into determination of the optimal fertilizing program, farmers may need help. Educational and demonstration programs conducted by departments of agriculture and agricultural extension services can inform farmers of the cost-effectiveness and energy-saving potential of the changes suggested here, as well as help the individual farmer work out the best approach to a particular combination of farming variables. Many farmers would appreciate cost-benefit comparisons of techniques or combinations of techniques

In conjunction with information transfer programs, states can conduct or fund research on new fertilizer timing and application methods. Nitrification inhibitors, for example, are relatively new, and more study is needed on how they affect soils, crops, and the environment; on what timing and rate of application yield optimum results; and on where these chemicals should be combined with other techniques for achieving more effective fertilizer use. Finally, states can conduct research to delineate the tradeoffs that will accompany shifts in timing and method of fertilizer application, such as from split application to spring application or to fall application with the addition of nitrification inhibitors. These studies can help predict whether potential energy savings will offset the expected costs.

Improved timing and methods of fertilizer application can mean substantial savings in energy and natural resources, can increase crop yields, and can enhance farmers' profits. These potential benefits impel states to take all possible steps to encourage the adoption of changes for more cost-effective use of fertilizer on the farm.

Reduce Dependence on Manufactured Nitrogen Fertilizers by Crop Rotation and Interplanting. Without nitrogen, crops cannot grow. Both farmers and agricultural researchers know that a lack of sufficient nitrogen is one of the major limiting factors to achieving

maximum crop yield. With the cost of manufactured nitrogen fertilizer so high, many farmers are looking for alternatives to provide the
needed nitrogen at less expense. Crop rotation and interplanting
(alternating regular crop rows with crops that add nitrogen to the
soil) can help farmers reduce their need for commercial nitrogen fertilizer. Because synthetic fertilizers are resource- and energy-intensive
as well as expensive, farmers who turn to alternative methods will
conserve both natural resources and energy as well as save money.

Farmers have been rotating crops for centuries. Long ago, they
noticed that planting one crop of corn after another or one crop of
cotton after another led to reduced yields. To maintain soil fertility,
they rotated their crops. The widespread use of commercially prepared fertilizers to prevent soil depletion is relatively recent. The
fertilizer price declines that made their use economically attractive
was, no doubt, a key factor in the decision of many farmers to
abandon crop rotation. By resurrecting crop rotation, farmers can
cut down on purchases of nitrogen fertilizers. For example, such
nitrogen-demanding crops as corn and cotton can be planted alter-
nately with a legume such as sweet clover. Legumes convert atmospheric nitrogen to forms that plants can use and reduce the need for
supplemental fertilizers. Planting clover and plowing it under a year
later puts nitrogen into the soil at a rate of about 150 pounds an
acre.[32] Often referred to as "green manure," this method also adds
valuable organic matter to the soil and, for corn, reduces disease and
weed problems. With crop prices high, however, most farmers are un-
willing to plant anything but cash crops.

Fortunately, one legume *is* a cash crop—the soybean. Corn and
soybeans can often be grown in rotation, saving 100 to 150 pounds
of nitrogen an acre.[33] Farmers can use the same machinery for both
crops and can shift easily from one crop to the other. In the Missis-
sippi delta, farmers can alternate cotton and soybeans and save 80
to 100 pounds of nitrogen an acre.[34] Of course, decisions to divert
acreage from one crop to another depend on crop prices and other
factors besides the nitrogen requirements of the crops.

Winter vetch and other versatile legumes can also "grow fertilizer
in place" when planted along with the regularly grown crop as cover
crops. An experiment in the northeastern United States has shown
that a cover crop planted in late summer and plowed under in early
spring adds 133 pounds of nitrogen an acre.[35] In the Southeast, a
cover crop could supply 70 to 80 pounds of nitrogen an acre.[36]
Since, however, winter cover crops do not combine well with corn,
this system of rotation is unsuitable for a large proportion of U.S.
farmland. Where cover crops are workable, however, the annual net

energy savings can amount to nearly 6 million BTUs an acre.[37] In addition to the energy savings, interplanting and plowing under a cover crop puts organic matter into the soil and reduces erosion of soil by wind and water, benefits that are difficult to quantify. Moreover, pollution of run-off water can be reduced or even eliminated when use of chemical nitrogen fertilizers is curtailed.

A new system of interplanting using crown vetch as the leguminous cover crop promises still greater benefits. Vetch is planted and allowed to grow for a year; herbicide sprays are then applied to weaken the vetch, and corn is planted by the no-till method. Growth of the young corn outstrips that of the weakened vetch, but the vetch eventually recovers and remains growing on the land after the corn is harvested. Come spring, the vetch is again treated with herbicides and a new corn crop planted. Preliminary tests of this system have resulted in annual savings of 100 pounds of nitrogen an acre.[38] If the crown vetch system proves practicable for half the acreage planted to corn in the United States, the estimated annual energy savings would be equivalent to over 70 trillion BTUs.[39] The crown vetch system works well for some soils, but at least one researcher questions whether this technique can lead to optimum crop production.[40]

Certain tradeoffs attend the transition from chemical fertilization to crop rotation and interplanting systems. Using "green manure" replaces the convenience of using chemical nitrogen fertilizer with greater demands on the farmer's time. Greater management skills are also required to rotate and interplant crops effectively. Thus, many farmers are understandably reluctant to make the switch.

To encourage farmers to rotate crops and to interplant with nitrogen-producing crops, states can sponsor information transfer programs conducted by personnel from their departments of agriculture and agricultural extension services. Farmers need estimates of the costs and benefits that particular methods of crop rotation and interplanting are likely to bring. Well-informed farmers will be better equipped to weigh the complex mix of benefits that various methods offer and to balance them with the costs.

States can also fund research and demonstration projects. Some of the methods suggested here for growing nitrogen fertilizers in place are not yet fully developed, and further research is required. Specifically, a particular technique must be tested for various soils and various climates; what works in the Corn Belt may not work elsewhere. Each state can concentrate on the development of those techniques that seem best suited to local conditions. Research programs might include scaled up farm demonstration projects. By supporting successful research on experimental farms, states can show cautious farmers that crop rotation and interplanting techniques are both practical and economical.

Farmers who modify their operations to incorporate some of these suggested changes deserve some economic protection. States can provide it through crop insurance programs, with a partial or total subsidization of the insurance premium. Participation in demonstration projects could be made a requirement of crop insurance programs. Successful projects will convince neighboring farmers that crop rotation and interplanting techniques are practical, economical alternatives to the use of energy-intensive and expensive commercial nitrogen fertilizers. Administered wisely, crop insurance programs protect farmers against crop failure and economic disaster as they encourage farmers to adopt new methods—all at little expense to the states.

Finally, states could consider taxing sales of commercially produced nitrogen fertilizers, based either on the selling price or on the energy content embodied in the finished product. This additional cost will encourage both substitution of "green manure" for chemical nitrogen fertilizer and increased production of crops that require relatively little nitrogen fertilizer. Before imposing such a tax, however, states must consider the tradeoffs. First, a state must ask whether it can offer its farmers a viable alternative to chemical fertilizers. If none of the techniques suggested in this section is suitable for the soil and climate of a particular locality, then a tax on nitrogen fertilizer will unfairly penalize farmers. Second, a state should consider whether taxes on fertilizers are practical. To what extent are farmers likely to dodge the tax by purchasing fertilizer from dealers in neighboring states? This practice could defeat the purpose of the fertilizer tax—energy conservation—and harm the state's economic well-being. In sum, taxes on fertilizer must be handled cautiously to avoid their becoming a large burden on farmers.

Supplement Chemical Fertilizers with Organic Wastes and Nutrient-Rich Run-off Water. Besides planting legumes that naturally enrich the nitrogen content of soil, farmers can apply waste materials—livestock manure, sewage sludge, and surface run-off water from livestock confinement areas and irrigated fields—to decrease their dependence on commercially produced fertilizers.[41] Among the benefits of using these materials are energy savings, reduced water pollution, improved soil condition, and diminished production costs.

Bulky animal manures and sewage sludge can take the place of commercial fertilizers, but there are tradeoffs: farmers must expect decreased convenience of application, heavier labor requirements, and additional capital investments.

Animal manure was probably the first material used by farmers to fertilize the soil. Today beef and dairy cattle, pigs, sheep, and chickens annually produce manure that weighs 1.7 billion tons when wet, and 340 million tons when dry.[42] Of this total, about 50 percent is produced in feedlot and confinement rearing facilities where recovery of the manure is possible; the other 50 percent is deposited widely on cropland, range, and pasture, where recovery is difficult and uneconomical.[43]

Animal manure today is considered a valuable supplement or replacement for commercially produced fertilizers, especially in light of the fertilizer price hikes that have occurred in recent years. Of the 50 percent of manure that is produced in confinement facilities where it is easily recovered, almost the entire amount is currently used by farmers as fertilizer.[44] Indeed, the demand has been so great that many Texas feedlot operators have had to backlog orders for manure.[45] Owing to the recent declines in the number of beef cattle held in concentrated production units, annual manure production is not likely to increase over the next few years. Possible energy savings are not easily calculated, and different studies have reached different conclusions. In 1975, the Environmental Protection Agency (EPA) estimated that natural fertilizer could have replaced 2.7 million tons of chemical fertilizers, or 5 percent of the total amount sold in 1973.[46] Projecting this figure into future annual energy savings, EPA estimated potential energy savings in 1977 at more than 42 trillion BTUs and for the period from 1977 to 1990 at 670 trillion BTUs.[47]

CAST has also estimated the energy-saving potential of using more animal manure, assuming that farmers could recover and use 15 percent more manure than they do now. Based on that assumption, it was estimated that annual savings could reach 30 trillion BTUs.[48]

Thus, manure not currently used as fertilizer could become a source of significant energy savings. Besides saving energy, the application of manure to cropland can lessen the severity of pollution problems that accompany manure production, storage, and disposal. For feedlot operators, the inevitable accumulation of manure can become an expensive headache: manure must be burned or hauled to landfills at great expense; water run-off from animal holding areas pollutes rivers and streams; and smoke from incinerators pollutes the air. Added to cropland soil, however, manure both beneficially augments the organic component (humus) and increases the number of bacteria in the soil. Thus enriched, soil is more easily tilled, better able to hold water, and less likely to erode.

Before deciding to substitute animal manure for commercial fertilizers, farmers should consider the possible drawbacks. First, manure is low grade fertilizer of varying composition. A much less concentrated source of plant nutrients than manufactured fertilizer, manure must be applied in large amounts; an average of ten tons an acre may be needed to produce optimal yields. Handling such large quantities of fertilizer means heavier labor, greater inconvenience, and additional investment in new equipment. Variations in the levels of plant nutrients, caused by different sources of the manure, complicate determination of the optimum rate of application. Soluble salts and trace elements present in animal manures may build up in the soil, harming crops and leaching into groundwater. Moreover, the availability to crops of nitrogen contained in manures depends upon the form the nutrient takes. Estimates vary as to the percentage of the total nitrogen content available for uptake by plants in the year after applying manure, but most fall short of 50 percent.

The sheer bulk of animal manure makes for troublesome transportation and storage. These drawbacks have compelled farmers to limit manure applications to cropland close to the source of manure production. Although this usage pattern is fine for a small, integrated operation that includes both livestock and crops, it is impractical for farmers who must import manure from feedlots located far from crop-growing areas.

Rises in the price of chemical fertilizer during the past few years have favorably changed the economics of transporting manure. Today, the USDA's Agricultural Research Service reports that shipping costs have become less formidable an obstacle to manure use than they were in the past.[49]

Storage presents additional problems, since animals produce manure continuously while crops need to be fertilized only at certain times of the year. To store the manure, expensive construction is necessary. Finally, the storage and use of manure creates inevitable odors that offend many people and may provoke public protest where farmlands abut populated areas. As the suburbs extend into farming areas this complaint is being heard more and more frequently.

Human wastes have long been used to fertilize crops in many parts of the world. Sewage sludge has come into use only recently in the United States, where the equivalent of 7 million tons of dry sewage sludge is produced annually.[50] Currently, only 20 percent of this amount is put onto the land, where it can be beneficial; the remainder is incinerated or dumped into landfills or into the ocean.[51] Thus, a fivefold increase in the use of sewage sludge for agriculturally productive uses is an immediate possibility.

The combined pressure of several forces is likely to impel states and municipalities to step up the use of sewage sludge on agricultural land. First, landfill capacity is dwindling rapidly. Second, strict air and water pollution controls are likely to curtail further incineration and ocean disposal of municipal and industrial wastes. Finally, compliance of municipal and industrial sewage treatment plants with the amendments to the Federal Water Pollution Control Act[52] is expected to multiply total sewage sludge production three to five times in coming years. CAST estimates that 20 to 35 million tons of dry sludge will become available annually, an amount roughly equivalent to the dry tonnage of manure likely to be gained through improved efficiency in manure recovery and use.[53]

Net energy savings resulting from the use of sewage sludge on cropland are not expected to be great. Based on the current level of sludge production, the 7 million tons of sewage sludge produced in a year can supply an estimated 0.2 percent of the nitrogen, 6.3 percent of the phosphorus, and 0.7 percent of the potassium supplied yearly by commercial fertilizers.[54] Even with the expected threefold to fivefold increase in sludge production, the contributions of nitrogen and potassium from sludge would still be negligible. Sludge could, however, replace a significant portion of the phosphorus now in use.

Applied to cropland, sludge, like manure, offers benefits in addition to energy savings. Some of the problems associated with conventional means of disposal can be eliminated. Many U.S. municipalities now incinerate sludge; if they all stopped now, air pollution would diminish and the equivalent of 50 million gallons of oil would be saved in the next year.[55] If the volume of sewage sludge grows as expected, annual oil savings could climb to between 1 billion and 1.75 billion gallons.[56] Sludge, like manure, also improves soil quality and helps prevent erosion.

Many of the same tradeoffs are involved whether sewage sludge or animal manure is used on cropland. Sewage sludge is a low grade fertilizer, with quantity and availability of plant nutrients variable; it is bulky; and its odor and other unpleasant associations may deter its use on food crops.

When sewage sludge is used to fertilize, contamination can pose a potential threat. Depending on the source, sludge can contain pathogenic bacteria, viruses, and parasites as well as heavy metals such as zinc, copper, nickel, and cadmium. All of these contaminants can endanger public health, and heavy metals are thought to hinder crop growth—and only little is known about many of these effects. Before farmers take up this system of fertilizing crops, agricultural policy-makers must learn more about the possible hazards of using sludge, so that meaningful guidelines and regulations can be issued. That farmers monitor the quality of their crops and water supplies will surely be required, and this will increase the costs of fertilizing with sludge.

Despite these drawbacks, the use of sewage sludge on farms is underway. The city of South Milwaukee, Wisconsin, ships its liquid sludge to farms sixty miles away, and both the city and the farmers find the arrangement economical. As fertilizer prices rise, transporting bulky sludge and manure over longer distances shows greater economic feasibility.[57]

Another source of some of the plant nutrients now provided by commercial fertilizers is run-off water. Rainfall and irrigation water inevitably wash nutrients away from crops, and many farmers deliberately overfertilize to allow for these expected losses. Unchecked, run-off from feedlots and livestock confinement areas on farms constitutes a significant source of water pollution. If farmers can control run-off, however, they will have at their disposal a rich source of irrigation water and of fertilizer. Captured and stored in small ponds, run-off water can replace commercial fertilizers to some extent, depending on its nitrogen level. Using aquaculture, farmers can also raise fish in these storage ponds, which can in turn be used as livestock feed or sold as a separate cash crop.[58]

How can states help farmers use natural wastes more effectively? Information transfer programs can help and should include information addressed to the general public as well as to farmers. First, agricultural extension services at the land-grant colleges and state departments of agriculture could join in devising programs to teach farmers about the advantages of using natural wastes. These efforts could include the latest findings about methods and rates of application, necessary management skills, and cost-effectiveness in substituting wastes for commercial fertilizers.[59] A second educational effort could seek to inform the general public about the social benefits of using organic wastes. The public needs assurance that the use of wastes in food production will not create health hazards. Anticipated in this way, public resistance to the use of sewage and other wastes can be minimized.

Without continuing research and development, educational efforts will be to little avail. To pin down potential health hazards, to eliminate odors, to develop more effective storage and application methods, and to stabilize inorganic forms of nitrogen for maximum fertilizer value, much research needs to be done. State research, development, and demonstration projects can promote more productive methods, reduce costs, increase convenience, and gain public acceptance of the use of wastes.

The use of sewage sludge offers potential benefits to farmers and society, but the risks must be understood. States should cautiously examine all available evidence in evaluating these risks before modifying any existing restrictions or drawing up new laws, regulations, and guidelines to encourage the use of sludge as fertilizer. State guidelines must be comprehensive in their protection of public health and of the environment, yet easy for farmers and regulators to understand. Farmers will be reluctant to invest in new equipment and storage facilities unless they clearly know how and when such fertilizers can be used. Wisconsin, for example, has set down guidelines for the application of waste water sludge to agricultural land.[60]

States could also control the quality of sludge at its source to enhance its value as fertilizer. Industries could be required to treat their wastes to remove traces of heavy metals before discharge into sewage systems. Of course, states must weigh the dollar costs of such a regulatory scheme against the benefits of reduced pollution and enhanced fertilizer quality.

States can back up educational programs and new guidelines and regulations with economic incentives. If farmers had insurance against economic loss, many might be willing to experiment with using sludge and manure as fertilizer supplements. States could subsidize crop insurance premiums for farmers who utilized manure and sludge.

Farmers' unwillingness to lay out large sums for new fertilizer-spreading equipment and storage facilities could frustrate efforts to encourage the increased use of manure and sludge. To offset this economic inhibition, states could offer state loans, loan guarantees, and tax incentives to farmers who invest in necessary equipment.[61]

Finally, states can help farmers measure the level of usable plant nutrients in their run-off collection ponds. By conducting water tests at little cost to farmers, states can help reduce operating costs, consumption of commercial fertilizer, and water pollution. Like soil tests, water tests can determine the presence or lack of nutrients necessary for crop growth. Provided by a state's department of agriculture or agricultural extension service, these tests are an inexpensive service that can result in substantial savings of both money and energy.

Conclusion

Fertilization of crops is essential to maintaining and improving farm productivity. The present heavy dependence of most U.S. farmers on commercial fertilizers makes fertilization the second most energy-intensive phase of food and fiber production, consuming over 500 trillion BTUs each year. For high crop yields, plants must have certain nutrients available to them at certain stages in their growth, but the nutrient source and method of application can vary. Matching fertilizer to soil conditions and plant needs, improving the timing and method of application, rotating crops to grow nitrogen in place, and supplementing commercial fertilizers with natural wastes all promise energy savings. Of course, the needs of soils, crops, and growing systems vary tremendously, and a technique good for one state may be totally inappropriate for another. Through a combination of selected strategies, policymakers can encourage farmers to make those changes suitable to local conditions. With a shared concern about conserving energy, money, and resources, states can help their farmers to change wasteful practices to help meet national goals at minimal sacrifice.

PESTICIDES

For most of the 9,000 years of agricultural history, farmers have controlled crop losses by relying on cultural practices and biological methods to fight pests.[62] Over the millenia, pests have been kept in check by tilling the soil, rotating crops, planting crops in particular combinations, controlling the application of fertilizer and irrigation water, and using benign predators. Around the beginning of the twentieth century, farmers turned to such simple chemical pesticides as kerosene, arsenates of lead, and nicotine. Only within the last thirty years have farmers come to rely on energy-intensive, complex chemical pesticides.[63] Most of the chemical pesticides used today are synthesized from petroleum, a nonrenewable energy source. This section of the chapter describes the extent of pesticide use in agriculture, how unnecessary pesticide use results in energy waste, and what can be done to reduce this waste.

Extent of Pesticide Use

The use of pesticides plays a significant role in the production of food and fiber. When judiciously used, herbicides, insecticides, rodenticides, and fungicides help maintain a plentiful supply of high quality food at low prices. Substitution of weed-killing herbicides

for one or more tillage operations can reduce the energy requirements for crop production (although the savings in tractor fuel are
partially offset by an increase in energy requirements for herbicide
production). The use of herbicides can also save water, especially in
arid regions, by reducing soil moisture loss that inevitably occurs
when farmland is tilled. A saving of water generally results in a saving of energy as well, since reduced pumping means reduced fuel
needs. Moreover, pesticide applications reduce losses from rodent
and insect contamination during food transportation and storage.

Despite these obvious benefits, the use of pesticides has sometimes
been accompanied by detrimental effects to human health and the
environment. To begin with, many pesticides have a potentially adverse impact on the environment because of their chemical and
physical properties. For example, some pesticides, such as the chlori-
nated hydrocarbons, are so chemically stable that they are not easily
broken down by natural forces or living organisms. Many pesticides
adhere to soil particles, allowing them to travel great distances and
to remain longer in the environment. Some pesticides are extremely
toxic to nontarget plants and animals. Other pesticides—compounds
of lead, arsenic, or mercury—can leave residues that may build up in
living organisms over many years. Still other pesticides are passed
along the food chain, from plants to animals to humans, with some
of the more persistent compounds accumulating in dangerous concentrations at the highest links in the food chain. Thus, pesticides
must be used cautiously to avoid the destruction of nontarget plants,
animals, humans, and the general environment.

Primary dependence on pesticides to control crop losses can even
be counterproductive in terms of crop production and pest control.
According to a lengthy report recently prepared by the National
Academy of Sciences, improper use of pesticides has led to a leveling
off, and in some cases a decline, in the production of corn and
cotton. Indeed, the report concludes with the prediction that crop
losses will accelerate in the future.[64]

To explain why crop production may decrease as pesticide use increases, it may be instructive to look at some of the inadequacies of
exclusive reliance on chemical insecticides. When farmers use insecticides, they often inadvertently kill helpful insects as well as target
pests. Harmless insects comprise about 99.9 percent of all insects and
include a great variety of natural predators that keep the harmful
insect population in check. To eradicate these helpful insect predators is to squander a cost-free natural method of controlling harmful
pests. The application of an insecticide often upsets the natural ecology of an area. Minor pests or formerly benign species may become a

cause of economic damage when the populations of their enemies or natural competitors are reduced or eliminated. Finally, a particular insecticide may become obsolete or ineffective when the target organism mutates and survives in the form of a resistant strain.

Pesticide application on American farms is already at a staggering level and is growing rapidly. David Pimentel and his colleagues at Cornell University studied the changes that have occurred in the production of the corn crop between 1945 and 1970. During this time, they report that the use of insecticides increased tenfold and the use of herbicides twentyfold.[65] The increased usage of pesticides has not been limited to the corn crop alone. Indeed, today nearly half of the insecticide used on crops is applied to cotton and tobacco.[66]

Looking at the actual quantity of pesticides used may also convey a feel for the rapid increase in their application. Between 1966 and 1971, for example, farmers increased their application of pesticides from 350 million pounds to almost 500 million pounds annually.[67] The annual increases since 1971 have been even greater. By 1974, it was estimated that farmers applied approximately 800 million pounds of pesticides to their crops, an annual increase of 100 million pounds.[68] And pesticide use continues to grow: although the final figures are not yet available, the U.S. Department of Agriculture estimates that pesticide demand in 1976 totaled approximately 1 billion pounds.[69]

The extent of acreage treated with pesticides has also increased, from 36 percent of all cropland in 1966 to over 50 percent in 1971.[70] Since 1971, the percentage of treated acreage has continued its upward climb, as farmers have placed more acres back into production and have increased their application of all pesticides, especially herbicides.

To produce this quantity of pesticides requires a great deal of energy-laden fossil fuel, primarily petroleum products for feedstocks and natural gas for heat and power. Since the chemicals used in the manufacture of pesticides are frequently refinery by-products, it is not an easy task to account for the energy content embodied in pesticides. Despite these difficulties, it is estimated that the energy required to produce the pesticides applied by farmers in 1971 was over 26 trillion BTUs.[71] Taking into account the 60 percent increase in pesticide use from 1971 to 1974 and the current annual increase of 5 to 10 percent, a good estimate of the energy devoted to pesticide application today is probably around 50 trillion BTUs.

In addition to the energy consumed in pesticide production, energy is needed to apply the pesticides. Spraying and crop dusting are major operations on many farms, calling for large quantities of fuel.

Some farmers are convinced that a particular crop must be sprayed at certain intervals throughout the growing season, sometimes as often as once a week. It takes little imagination to grasp the consequences of this practice when multiplied by the many farmers who apply pesticides by force of habit.

Although the energy expended to produce these crop-protecting pesticides amounts to only about 1 percent of all energy used in agricultural production and mere tenths of a percent of national energy consumption, energy is wasted by unnecessary pesticide use.[72] Reducing this unnecessary use through changes in policy and in methods of pest control can result in substantial savings of labor, energy, and money without reducing quality or quantity of crop yield. All told, it is estimated that 35 to 50 percent of all pesticides used and energy expended could be conserved.[73]

Problems of Pesticide Use

In the face of resource shortages and production cost squeezes, farmers must be careful to use pesticides only when absolutely necessary. All too often, however, energy-intensive chemical pesticides are overused. In this section, "unnecessary pesticide use" means any application of pesticides where the dollar costs exceed the dollar value added or where a more judicious pest control action could have been selected.

Energy is wasted when farmers apply a pesticide

- Where a need has not been established,
- Where cultural practices are feasible alternatives,
- Under improper weather conditions,
- That is wrong for the job,
- At a time when the target pest is not vulnerable, or
- Inaccurately through faulty spraying technique, calibration, or target selection.

Such errors result in the loss of resources and money. To minimize losses and maximize profits, farmers must carefully plan their pest control program.

The first step calls for a reevaluation of all sources of information to determine their accuracy and reliability. There are many sources of pest management information, and the usefulness of each source varies, as farmers well know.[74] For example, a frequent source of pest control information is farm magazines. In a recent study, however, farmers rated that information source as one of the least useful. The same study also found that some of the most useful information sources—including county agents and university extension specialists —were consulted only infrequently.[75]

Many of the information sources are biased in favor of continued reliance on chemical pesticides. The pesticide industry certainly has a stake in presenting information in the best light possible, and thus the claims of pesticide salesmen and farm magazine advertisements should be discounted to some extent. Friends and neighbors are also important sources of pest control information and may sometimes perpetuate current practices. The social pressure exerted by friends may sometimes influence a farmer to do what others are doing, even when a different pest control method would be more beneficial.

Getting useful information to farmers about new pest management techniques need not be a difficult task. The U.S. Department of Agriculture and the land-grant colleges are the logical choices for developing information transfer programs. A network of Cooperative Extension Service pest management specialists and county agents is in place, and farmers perceive these information sources as useful. There are, however, two disadvantages. Perhaps the more important consideration is financial; setting up a program to provide pest management alternatives will often require more specialists or funds than are available. The other problem is institutional. In the past the USDA has shown ambivalence, modestly supporting research about pest control alternatives that rely less heavily on chemicals while actively promoting the continued use of pesticides.

The second step in planning a pest control program is to have farmers reassess their own attitudes toward pesticide use. Pesticides are popular for many reasons: they are convenient to use, relatively inexpensive, and generally effective. Understandably, many farmers are content to continue using the methods that have worked in the past. Perhaps there is a fear that innovative pest control measures might lead to crop yield reductions, and therefore, a change to alternative methods is resisted. Before farmers will adopt methods that rely less heavily on chemical pesticides, they must be convinced that the new techniques will work. The goal of any pest control program is to minimize economic losses caused by pests, that is, pest control. Application of additional pesticides will eliminate more pests, but this may be uneconomic; the cost of such overkill techniques may be more than the value of the crops that these pests would have ruined. It is estimated that treating crops only when absolutely necessary rather than on a routine schedule could result in an annual savings of over 12 trillion BTUs.[76]

The third step in the reevaluation process requires growers, buyers, and policymakers to work together to eliminate private and public legal constraints that may lead to unnecessary pesticide use. For example, private contractual arrangements may cause waste. Food processors and wholesalers generally rely on contracts with growers to guarantee delivery of an adequate supply of high quality produce. Buyers sometimes specify, as one of the terms of the contract, that growers use particular pesticides according to a schedule. Contracts made in advance of the growing season may make pesticide application necessary, even if no pest problems arise.

In addition, some food processors and wholesalers demand that farmers meet food quality standards in excess of the standards set by the state and federal governments. Government regulations that strive to protect the public health and to increase the marketability of produce are a boon to consumers and growers alike. But private standards that exceed government standards may influence growers to use pesticides as crop insurance even when no pest problem exists.

Finally, private marketing orders are a third type of legal constraint that may result in unnecessary pesticide use. In California, a major agricultural state, food producers and handlers are allowed to establish marketing order advisory boards. With the approval of the state department of agriculture, these boards can regulate food production and marketing in the state.[77] The marketing orders, which are intended to stabilize the income of farmers, may result in unnecessary pesticide application if the local board sets overly stringent quality standards. Moreover, the board has the power to order the destruction of produce already grown, a step guaranteed to waste all the energy-intensive resources that were used to grow the crop.

Techniques and Strategies for Reducing Unnecessary Pesticide Use

Despite growing evidence that increased pesticide application results in diminishing economic returns, in energy inefficiency, and in damage to the environment, many farmers are still firmly committed to the unbridled use of pesticides. This should come as no surprise. During the past three decades, farmers have embraced many new agricultural chemicals that seemed to offer miracle solutions to farm problems. The use of chemical pesticides has become an accepted practice with many farmers, a habit that they may be reluctant to change. One reason why new pesticide measures have not been readily adopted is that fully effective and practical alternatives are either unavailable or unknown to growers in many parts of the country.

In fact, many successful programs of pest management have been developed that can at once reduce pesticide application, decrease energy use, and maintain or increase profits. These programs, each of which combines various pest control practices, are known as "integrated pest management."[78] Integrated pest management is an approach that maximizes natural control of pest populations. Based upon knowledge of the crop, of pests and their natural enemies, and of the environment, farming practices are modified to affect potential pests adversely and to aid natural enemies of the pests.

After these initial preventive steps are taken, the fields are monitored to determine the level of pests, their natural enemies, and other important environmental factors. Suppressive measures are initiated only when economically significant crop damage from the pest is determined to be likely. And when suppressive measures are required, pest control is achieved by using techniaues that will cause minimal disruption to the pests' natural enemies. These steps can include the use of biological controls, insect pathogens, and selective spraying with pesticides.

Some pest management methods, which are today seen as alternative cultural practices, are as old as agriculture itself. One effective cultural form of pest control practice is crop rotation. By changing from a pattern of continuous corn, for example, farmers can deprive pests such as the corn earworm of an established home and a source of nutrition. Today's monoculture farming, however, does not lend itself to this practice, since most farmers tend to concentrate on only one or perhaps two crops.

A related practice that retards or discourages the spread of pests is the planting of crops in particular combinations. Some farmers plant "trap crops"—that is, crops grown to lure inspect pests away from the main crop. For example, alfalfa is sometimes planted near cotton to draw harmful insect pests away from the valuable cotton crop. Some farmers also combat pests by providing a hospitable environment for the pests' natural predators. Some California grape growers plant evergreen blackberry bushes near their vineyards to provide a winter home for the parasitic wasp that controls one insect pest. And it is common practice for organic gardeners to intersperse their crops with herbs that repel unwanted insects.

Other pest management techniques are more recent developments. Many farmers now use new crop varieties that are more disease resistant than older varieties. Without the great strides that have been made in genetic methods for increasing disease resistance, the current high levels of production would not be possible. Moreover, the development of biological controls has advanced significantly. In addition to the traditional use of some types of predators and parasites, scientists have found how to make more effective use of insect pathogens and insect attractants in suppressing pests. Finally, the release of sexually sterile insects has been used to keep some pest populations in check.

In the broadest sense, integrated pest management encompasses the entire spectrum of pest control practices, with the emphasis on nonchemical means where feasible. This approach differs significantly from the now common application of pesticides on a fixed schedule. The integrated approach requires early detection of pests, identification of the threshold level of pest population causing economic harm, and deployment of a combination of measures as needed. In this way, pest damage is limited, energy conserved, and the environment preserved.

Monitor Pests.   The heart of an integrated system of pest management is the early detection of pest populations in a particular area. Most often, field scouts are hired to determine the population levels of pests and their natural enemies. Having this information at an early stage gives farmers the opportunity to weigh alternative courses of action and to apply pesticides only when absolutely necessary, rather than on a routine fixed schedule. Joint federal and state efforts, including monitoring programs, have been carried on in many states. In 1971, for example, the Cooperative Extension Service initiated two pilot test programs, one for tobacco in North Carolina and one for cotton in Arizona.[79] Since that time, the program has been expanded to include thirty-nine projects in twenty-nine states for nineteen different crops.[80]

One way that states can encourage this type of program is by subsidizing the scouts' salaries for a period of time, perhaps three years, after which the farmers would bear the entire cost. Besides reducing pesticide use and enhancing energy conservation efforts, this program could generate employment opportunities for private crop protection specialists.

A more advanced step is to program computers to track the spread of pests in an area, helping specialists determine if and when pesticide applications are necessary. Researchers at Michigan State University, for example, have developed a computerized pest management system that is used to furnish the most recent information to all Michigan farmers via a telephone advisory network.[81] According to Dean Haynes, the Michigan State University entomologist who helped devise the system, the results have been heartening. To cite one example, the quantity of miticides used by apple growers has dropped to only one-seventh the amount in use prior to the program.[82] Other states could develop similar computer-based programs for their major crops, using the facilities of their land-grant college or university.

Step Up Research and Development of Alternative Pest Control Methods. To expand the available information about alternative methods of pest control, states can fund research, development, and demonstration projects that show how farmers can rely less heavily on pesticides. The cost of funding such programs need not be borne entirely by the states. The U.S. Department of Agriculture has been conducting and sponsoring research and field tests in the area of integrated pest management for many years and is continuing its efforts to develop new pest management techniques. Through the Cooperative Extension Service, the Cooperative State Research Service, the State Agricultural Experiment Stations, and the land-grant colleges, states have participated actively in developing these techniques. Joint state and federal programs could further develop alternative methods of pest control.

Moreover, the Federal Insecticide, Fungicide, and Rodenticide Act (FIFRA) provides that the administrator of the Environmental Protection Agency, in cooperation with other federal agencies, universities, and others "shall give priority to research to develop biologically integrated alternatives for pest control."[83] States could, through their land-grant colleges, devise programs to research the feasibility of alternative pest control methods and apply to EPA for federal funding to supplement state funds.

State-financed efforts in California and Hawaii have produced some of the most valuable advances in biological control research,[84] and additional state funding of agricultural research can augment the knowledge already obtained.

Increase Funding for Education and Information Programs. An effort to increase farmers' knowledge of integrated pest management techniques is an important element in any program to reduce the current reliance on chemical pesticides. Most farmers today accept pesticides as the weapon of choice in controlling pests, but that attitude can be changed. States must provide farmers with sound information if energy-conserving programs in crop protection are to succeed. The land-grant colleges, university extension specialists, and county agents all are useful sources of information that can help farmers reduce pesticide application while maintaining productivity. Although farmers consider these information sources very useful, farmers most frequently turn to other sources that generally favor pesticide use.[85]

Some states are very active in developing programs of biological pest control and in disseminating useful information to farmers. The most useful first step is for each state to assess its present system of delivering information about biological pest control methods to its farmers. Some states may find that their integrated pest management efforts have not been fully developed; others may find that farmers use the programs infrequently because of a lack of specialists with information about integrated pest management. Those states could improve their education and information programs by providing additional funds for more extension specialists or county agents. Additional funding will permit an increased number of classes, demonstrations, and individual consultations.

Certification of Pesticide Users.   FIFRA requires the certification of those who are going to use certain pesticide products, in order to protect the safety of the user and the public and to reduce damage to the environment.[86] These amendments specify that pesticides must be grouped into two classes: those for "general" use and those for "restricted" use. Pesticides classified in the latter category are subject to controls in addition to what is listed on the container label, because they are deemed to pose special hazards to the user or the environment. Restricted pesticides will be available only to certified pesticide applicators, including both custom pesticide applicators and private applicators.

The certification process is largely a responsibility of the states, which are required to develop training programs for applicators who will use restricted pesticides. FIFRA mandates that these state programs be based on and conform to federal standards set by the EPA administrator before they receive approval.[87] Some of the bite has been removed from the certification program, however, since the statute also says that state training programs can "not require the private applicators to take, pursuant to a requirement prescribed by the Administrator, any examination to establish competency in the use of the pesticide."[88] In promulgating regulations pursuant to the statute, the EPA administrator requires states that are planning to certify pesticide applicators to stress a number of subjects in their training programs, including recognition of pest types, knowledge of pest damage, and methods of properly applying pesticides.[89]

In the process of certifying pesticide applicators, states have the flexibility to initiate programs that encourage energy conservation. For example, state training programs can include instruction in how to apply pesticides more prudently in performing a particular job. The FIFRA amendments of 1975 use the certification procedure as a way to get states to offer instruction in integrated pest management techniques. By law, the standards prescribed by the EPA administrator for the certification of pesticide applicators and state plans submitted to the administrator must include provisions for teaching individuals about integrated pest management techniques. These plans, however, may not require individuals to receive such instruction nor may they require individuals to display competence in using integrated pest management techniques.[90]

Thus, certification programs go hand in hand with other educational efforts. Teaching farmers how to apply pesticides most efficiently is one way that a state can use the certification procedure to encourage energy conservation.

Modify Legal Constraints. Private legal restrictions can increase the total amount of pesticide used by farmers. Some states have, by statute, empowered groups within the private sector to set food production and marketing orders that can lead to pesticide waste. One result of excess pesticide use is energy waste, and a state can improve energy efficiency in agriculture by modifying laws that permit energy waste to occur. Thus, if studies show that private marketing orders lead to energy waste, a state could amend the statute delegating authority to issue those marketing orders.

The difficulties facing those who would amend these marketing laws, however, are formidable. The grower groups, which have been delegated the authority to regulate production and price, can be expected to fight any suggested changes that would reduce their power. And the political realities are such as to make changes in these laws unlikely.

Conclusion

Over the last twenty-five years, many farmers have come to rely on chemical means to control crop pests. Often, this has resulted in the unnecessary application of pesticides that can adversely affect the environment, pose a threat to human health, increase a farmer's costs of production, and squander limited resources. Implementation of some of the previously discussed strategies can provide three important benefits to farmers and everyone else.

First, the suggested changes can cut energy consumption significantly, perhaps in half, freeing energy for other uses. Second, the changes can enable farmers to keep pests in check at no increase in cost, thereby allowing farmers to maintain or improve profits. Third, and perhaps most important, the risk of harm to humans and the environment from highly toxic chemical pesticides can be reduced.

The full impact on the earth and its inhabitants of such long-lasting pesticides as DDT, dieldrin, and aldrin is only now being understood. A greater appreciation of all the effects of pesticides may be the key to reduced pesticide use in the future.

NOTES TO CHAPTER 4

1. David Pimentel et al., "Food Production and the Energy Crisis," *Science* 182 (November 2, 1973): 445; Gary Heichel, *Comparative Efficiency of Energy Use in Crop Production*, Bulletin no. 739 (New Haven: Connecticut Agricultural Experiment Station, 1973), p. 18.

2. The energy expenditures in the last two categories are sometimes combined into one figure, representing all indirect consumption of energy in farm production. *See, e.g.*, Eric Hirst, "Food-Related Energy Requirements," *Science* 184 (April 12, 1974): 136. Hirst estimates that 44 percent of the total energy used in agriculture is consumed directly and 56 percent indirectly.

3. Arriving at a precise percentage or figure is not easy. In an early study based on 1963 data, Hirst concludes that the production of indirect and capital inputs consumed 1,230 trillion BTUs or 2.5 percent of all energy used in 1963. Hirst, *supra* note 2 at 136. The National Academy of Sciences estimated that the energy consumed in producing indirect and capital inputs was 1.9 percent of all energy used in 1972. Using an estimate of 72.3 quadrillion BTUs for national energy consumption in 1972, this means that approximately 1,370 trillion BTUs were consumed to produce these inputs. National Academy of Sciences, *Agricultural Production Efficiency* (Washington, D.C.: U.S. Government Printing Office, 1974), p. 9. A current estimate, prepared for the Federal Energy Administration, puts indirect energy consumption for the production of fertilizers, pesticides, and other indirect inputs at 750 trillion BTUs (1.1 percent of total national energy use) and for the production of capital inputs at 275 trillion BTUs (0.4 percent of total national energy usage). This estimate may be low, due to the limited scope of the study. The report calculates the energy used for domestic food production only and excludes the energy used to produce non-food crops (such as cotton and tobacco) and exported food and fiber. Adding the indirect energy used to produce the excluded crops would double the figures. Booz, Allen & Hamilton, Inc., *Energy Use in the Food System*, prepared for the Federal Energy Administration (Washington, D.C.: U.S. Government Printing Office, 1976), pp. IV−2, IV−7, IV−8, IV−19, IV−20.

4. U.S. Department of Agriculture, Economic Research Service, *The U.S. Food and Fiber Sector: Energy Use and Outlook*, prepared for the U.S. Senate Committee on Agriculture and Forestry (Washington, D.C.: U.S. Government Printing Office, 1974), p. 4.

5. John S. Steinhart and Carol E. Steinhart, "Energy Use in the U.S. Food System," *Science* 184 (April 19, 1974): 309.

6. Arthur D. Little, Inc., *Economic Impact of Shortages on the Fertilizer Industry*, prepared for the Federal Energy Administration (Springfield, Virginia: U.S. Department of Commerce, National Technical Information Service, 1975), p. III–2.

7. Denis Hayes, *Energy: The Case for Conservation*, Worldwatch Paper 4 (Washington, D.C.: Worldwatch Institute, 1976), p. 41.

8. Pimentel et al., *supra* note 1 at 444–45.

9. U.S. Department of Agriculture, Economic Research Service, *Changes in Farm Production and Efficiency*, Statistical Bulletin no. 561 (Washington, D.C.: U.S. Government Printing Office, 1976), p. 25.

10. U.S. Department of Agriculture, Statistical Reporting Service, *Crop Production*, Publication CrPr 2–2 (Washington, D.C.: U.S. Government Printing Office, October 1976), p. A–3; Council on Environmental Quality, *Sixth Annual Report* (Washington, D.C.: U.S. Government Printing Office, 1975), pp. 456–57.

11. Council for Agricultural Science and Technology, *Fertilizer Practices and Efficiency of Use*, CAST Report no. 37 (Ames: Iowa State University, 1975), p. 3.

12. Council for Agricultural Science and Technology, *Potential for Energy Conservation in Agricultural Production*, CAST Report no. 40 (Ames: Iowa State University, 1975), p. 14.

13. *Ibid.*

14. *Ibid.*

15. CAST Report no. 37, *supra* note 11 at 9.

16. Arthur D. Little, Inc., *supra* note 6 at I–32 to I–35; U.S. Department of Agriculture, Statistical Reporting Service, *Agricultural Prices Annual Summary 1975*, Publication Pr 1–3 (Washington, D.C.: U.S. Government Printing Office, June 1976), p. 156.

17. Arthur D. Little, Inc., *supra* note 6 at I–32.

18. *Changes in Farm Production and Efficiency*, *supra* note 9 at 2, 25.

19. Lester R. Brown and Erik P. Eckholm, *By Bread Alone* (New York: Praeger, 1974), p. 118.

20. Arthur D. Little, Inc., *supra* note 6 at I–83 to I–88.

21. CAST Report no. 37, *supra* note 11 at 7.

22. Edward Groth III, "Increasing the Harvest," *Environment* 17, no. 1 (January–February 1975): 32.

23. The U.S. Department of Agriculture and Federal Energy Administration have recently published a series of energy-saving guides for farmers that include specific suggestions for reducing energy consumption in fertilizer use. Allen Schienbein, *A Guide to Energy Savings for the Field Crops Producer*, prepared for the U.S. Department of Agriculture and the Federal Energy Administration (Washington, D.C.: U.S. Department of Agriculture, 1977), pp. 15–18; N.A. Wynn, *A Guide to Energy Savings for the Orchard Grower*, prepared for the U.S. Department of Agriculture and the Federal Energy Administration (Washington, D.C.: U.S. Department of Agriculture, 1977), pp. 8–10; N.A. Wynn, *A Guide to Energy Savings for the Vegetable Producer*, prepared for the U.S. Department of Agriculture and the Federal Energy Administration (Washington, D.C.: U.S. Department of Agriculture, 1977), pp. 8–12.

24. *See* Schienbein, *supra* note 23 at 16; Wynn, *A Guide to Energy Savings for the Orchard Grower, supra* note 23 at 9; Wynn, *A Guide to Energy Savings for the Vegetable Producer, supra* note 23 at 9.

25. C.G. Coble and W.A. LePori, *Energy Consumption, Conservation and Projected Needs for Texas Agriculture*, Publication S/D—12, Special Project B, prepared for the Governor's Energy Advisory Council (College Station: Texas Agricultural Experiment Station, 1974), p. 43. *See also* CAST Report no. 40, *supra* note 12 at 14.

26. CAST Report no. 40, *supra* note 12 at 15.

27. *Ibid.* This calculation was based on an application rate of 110 pounds of nitrogen an acre on 70 million acres of corn. The amount of nitrogen applied to corn has decreased somewhat, and the estimate is therefore slightly high.

28. D.W. Nelson et al., "Conserving Energy with Nitrification Inhibitors," in *Agriculture and Energy*, ed. William Lockeretz (New York: Academic Press, 1977), pp. 361, 369—71.

29. *Ibid.*, pp. 373—74.

30. *Ibid.*

31. *Ibid.*, pp. 374—75.

32. Pimentel et al., *supra* note 1 at 446.

33. CAST Report no. 37, *supra* note 11 at 14.

34. *Ibid.*

35. Pimentel et al., *supra* note 1 at 447.

36. CAST Report no. 37, *supra* note 11 at 14.

37. *See, e.g.*, Pimentel et al., *supra* note 1 at 447; Steinhart and Steinhart, *supra* note 5 at 313. This figure must be deemed a high estimate of the potential energy savings, since farmers generally do not apply 133 pounds of nitrogen to their crops.

38. CAST Report no. 40, *supra* note 12 at 16.

39. *Ibid.*

40. Communication between R.G. Hoeft, assistant professor of soil fertility, University of Illinois College of Agriculture, and the author, March 18, 1976.

41. A study of the practicability, desirability, and feasibility of using organic waste materials must be prepared by the secretary of agriculture by September 29, 1978. *See* the Food and Agriculture Act of 1977, 7 U.S.C.A. § 3303 (West Supp. February 1978).

42. Council for Agricultural Science and Technology, *Utilization of Animal Manures and Sewage Sludges in Food and Fiber Production*, CAST Report no. 41 (Ames: Iowa State University, 1975), p. 6.

43. *Ibid.*

44. *Ibid.*

45. Coble and LePori, *supra* note 25 at 40.

46. Mathematica, Inc., and Peat, Marwick, Mitchell and Co., *Comprehensive Evaluation of Energy Conservation Measures*, prepared for the U.S. Environmental Protection Agency (Washington, D.C.: U.S. Government Printing Office, 1975), p. III—85.

47. *Ibid.*, p. III—86.

48. CAST Report no. 41, *supra* note 42 at 6, 19.

49. Mathematica et al., *supra* note 46 at III—84.

50. CAST Report no. 41, *supra* note 42 at 5.

51. *Ibid.*

52. 33 U.S.C. § 1251–1376 (Supp. V 1976 & West U.S.C.A. Supp. February 1978). *See also* Council on Environmental Quality, *Sixth Annual Report*, *supra* note 10 at 77—78, 82.

53. CAST Report no. 41, *supra* note 42 at 6—7.

54. *Ibid.*, p. 7.

55. *Ibid.*, p. 21.

56. *Ibid.*

57. Jane Brody, "Wisconsin Town Uses Its Waste to Help Grow Crops, Creating a Model for the State," *New York Times*, July 9, 1976, p. A—9.

58. The development of aquaculture is encouraged by the Food and Agriculture Act of 1977. *See* 7 U.S.C.A. § 1932(a) (West Supp. February 1978).

59. *See, e.g.*, William Lockeretz et al., "Economic and Energy Comparison of Crop Production on Organic and Conventional Farms," in *Agriculture and Energy*, ed. William Lockeretz (New York: Academic Press, 1977), pp. 85—101.

60. Dennis R. Keeney, Kwang W. Lee, and Leo M. Walsh, *Guidelines for the Application of Wastewater Sludge to Agricultural Land in Wisconsin*, Technical Bulletin no. 88 (Madison: Wisconsin Department of Natural Resources, 1975).

61. For example, the secretary of agriculture has the power to make loans and loan guarantees to farmers who conserve, develop, and utilize aquaculture. 7 U.S.C.A. § 1932(a) (West Supp. February 1978). In addition, loans may be made to help farmers purchase farm equipment that utilizes solar energy. 7 U.S.C.A. § 1942(a)(2) (West Supp. February 1978).

62. The term "pest" means any insect, rodent, nematode, fungus, weed, or any other form of plant or animal life which the administrator of the Environmental Protection Agency declares to be a pest. *See* the Federal Insecticide, Fungicide, and Rodenticide Act, 7 U.S.C. § 136(t) (Supp. V 1976).

63. The term "pesticide" means any substance or mixture of substances intended for preventing, destroying, repelling, or mitigating any pest, and any substance or mixture of substances intended for use as a plant regulator, defoliant, or desiccant. *See* the Federal Insecticide, Fungicide, and Rodenticide Act, 7 U.S.C. § 136(u) (Supp. V 1976). Thus, the all-inclusive term pesticide refers to herbicides, insecticides, fungicides, and other miscellaneous chemicals.

64. National Academy of Sciences, *Pest Control: An Assessment of Present and Alternative Technologies*, 5 vols. (Washington, D.C.: National Academy of Sciences, 1975—76), 1:17.

65. Pimentel et al., *supra* note 1 at 444.

66. David Pimentel, "Extent of Pesticide Use, Food Supply, and Pollution," *J. of N.Y. Entom. Soc.* 81 (March 1973): 15.

67. Paul A. Andrilenas, *Farmers' Use of Pesticides in 1971 ... Quantities*, prepared for the U.S. Department of Agriculture, Economic Research Service, Agricultural Economic Report no. 252 (Washington, D.C.: U.S. Government Printing Office, 1974), pp. 4—6.

68. D. Lee Fowler and John N. Mahon, *The Pesticide Review 1975*, prepared for the U.S. Department of Agriculture, Agricultural Stabilization and

Conservation Service, Agr. Stab. and Cons. Serv. Rep. no. 155 (Washington, D.C.: U.S. Government Printing Office, 1976), pp. iii, 1, 7.

69. Paul A. Andrilenas and Theodore R. Eichers, *Evaluation of Pesticide Supplies and Demand for 1976*, prepared for the U.S. Department of Agriculture, Economic Research Service, Agricultural Economic Report no. 332 (Washington, D.C.: U.S. Government Printing Office, 1976), pp. v, vi.

70. Paul A. Andrilenas, *Farmers' Use of Pesticides in 1971 . . . Extent of Crop Use*, prepared for the U.S. Department of Agriculture, Economic Research Service, Agricultural Economic Report no. 268 (Washington, D.C.: U.S. Government Printing Office, 1975), p. 3.

71. Albert Fritsch et al., *Energy and Food*, CSPI Energy Series VI (Washington, D.C.: Center for Science in the Public Interest, 1975), p. 7. A similar estimate is offered by David Pimentel and his co-workers at Cornell University, who calculated the energy consumed in treating the corn crop with herbicides and insecticides in 1970. Extrapolating from the data that they have compiled, the energy needed to treat all crops with these pesticides in 1970 was more than 23 trillion BTUs. *See* Pimentel et al., *supra* note 1 at 444–45.

72. *The U.S. Food and Fiber Sector, supra* note 4 at xiv–xv.

73. Steinhart and Steinhart, *supra* note 5 at 313.

74. Rosemarie von Rumker et al., *Farmers Pesticide Use Decisions and Attitudes on Alternate Crop Protection Methods*, prepared for the U.S. Environmental Protection Agency (Washington, D.C.: U.S. Government Printing Office, 1974), pp. 107–14.

75. *Ibid.*, p. 111.

76. CAST Report no. 40, *supra* note 12 at 14.

77. California Marketing Act of 1937, Cal. Food & Agric. Code §§ 58741–58748 (West 1968).

78. *See, e.g.*, Council on Environmental Quality, *Integrated Pest Management* (Washington, D.C.: U.S. Government Printing Office, 1972).

79. *Ibid.*, pp. 32–33. *See also* National Academy of Sciences, *Pest Control, supra* note 64 at 1:214.

80. National Academy of Sciences, *Pest Control, supra* note 64 at 1:214–16.

81. *See* "Michigan Farmers Use Phone for Advice on Spraying," *New York Times*, August 1, 1976, p. 22.

82. *Ibid.*

83. 7 U.S.C. § 136r(a) (Supp. V 1976).

84. Communication between Dr. Elinor Cruze Terhune, Office of Policy Research and Analysis, National Science Foundation, and the author, October 18, 1976.

85. *See* von Rumker et al., *supra* note 74 at 111.

86. 7 U.S.C. § 136b (Supp. V 1976); 40 C.F.R. § 171.7 (1976).

87. 7 U.S.C. § 136b (Supp. V 1976).

88. 7 U.S.C. § 136b(a)(1) (Supp. V 1976).

89. 40 C.F.R. §§ 171.4, 171.5 (1976).

90. 7 U.S.C. § 136b(c) (Supp. V 1976).

Chapter 5

Meat, Poultry, and Dairy Production

INTRODUCTION

Any discussion of the production of beef cattle, hogs, poultry, and dairy cows must begin with a clear understanding that most Americans like to eat meat, eggs, and milk. Consumer demand for livestock products, especially beef, remains strong. Despite the recent instability in price and demand, it is still expected that Americans will increase their consumption of beef from 129 pounds per person in 1976 to 158 pounds per person by the end of this century.[1]

Policymakers must also consider that livestock are valuable as food conversion machines. Beef cattle, hogs, and sheep are able to convert low grade roughages and food-processing wastes into high quality food products that are excellent sources of protein and food energy. Indeed, increasing affluence and changing consumer demand now results in the average American deriving about 40 percent of his daily caloric intake from these sources.[2]

At the same time, the current methods of livestock production have their price. Placing beef, bacon, and broilers on the dining room table puts a large demand on the limited supply of agricultural resources—tillable land, petroleum fuels, water, and fertilizers—leading many serious observers of American agriculture to question the wisdom of how these resources are utilized. Both critics and defenders of the present methods agree on one point: raising animals is a net energy loser.

While there is some dispute over whether the energy efficiency of production is improving or declining, the stronger philosophical disagreement is about whether to continue our current practices, particularly feeding grain to beef cattle. The harshest critics argue that such production is morally unjustifiable in a world that is unable to feed all its people adequately. If animals ate only crop wastes or grasses that could not be consumed by people, little criticism would be heard. But animals are fed grain that embodies large amounts of energy and other resources; cattle are grazed on pasture that has been fertilized and irrigated; and energy-intensive, high protein feed supplements are added to produce rapid weight gain. Moreover, livestock are housed in buildings that are heated and ventilated using large quantities of fuel and electricity. Since energy conservation is a national objective with the highest priority, states should accord the same priority to the development of strategies that will reduce energy consumption in livestock production.

PRESENT LEVEL OF ENERGY USE

To properly frame the discussion, one must consider that a staggering 88 percent of the annual food harvest is first fed to animals; only 12 percent of the harvested crop goes directly into our mouths without first having been chewed and digested by livestock.[3] The beef cattle industry, which is the largest single component of American agriculture, offers a good example of the extensive use of resources in livestock production.

To obtain the 15 billion pounds of retail cuts of beef that were carved from 34 million head of cattle slaughtered in 1973, livestock ranchers used a combination of feeding methods, running the gamut from range grazing to feedlot finishing with many variations in between.[4] Unlike the early days of beef production, when most cattle grazed on available pasture and range until they were ready for slaughter, beef cattle today are fed large amounts of grain. Since World War II, cattle ranchers have found it profitable to bring calves to final market weight in feedlots on a diet of harvested feed grain. This system is now so prevalent that beef cattle consume about 50 million tons of grain (primarily corn) each year.[5]

The enormous quantities of grain used for feed translate into an astounding use of energy. According to a report recently prepared for the Federal Energy Administration, an estimated 225 trillion BTUs were consumed in the production of all livestock for domestic consumption in 1973.[6] The direct energy consumption was divided among livestock producers as follows:

Beef cattle	92 trillion BTUs
Dairy cows	52 trillion BTUs
Poultry	40 trillion BTUs
Hogs	37 trillion BTUs
Sheep	4 trillion BTUs[7]

When one adds to this amount the energy used indirectly to grow and process feed for these animals, the total energy consumption is vastly greater.

Livestock Production Is a Net Energy Loser

Many critics who express alarm at the large commitment of energy and other resources to livestock production are actually concerned about the inefficiency with which livestock convert feed into food. For example, Heichel and Frink report that it takes one hundred calories of feed to produce nineteen calories of dairy goods, thirteen calories of pork, twelve calories of poultry, or a mere five calories of beef, veal, or lamb.[8]

William Splinter, an agricultural engineer at the University of Nebraska, has calculated the conversion efficiency in a different way, focusing on the amount of feed grain that is required to produce a pound of dressed meat. He concludes that the feedlot method for fattening beef is the least efficient method of converting feed into meat: 34 pounds of feed grain are needed to produce a pound of beef; 15 pounds of feed grain to produce one pound of pork; and 9.7 pounds of feed grain for each pound of chicken.[9]

David Pimentel and his colleagues at Cornell University concentrated on yet another measure of production efficiency, the amount of protein produced by each animal. The livestock that most efficiently converted feed into protein were broiler chickens, followed by egg-laying hens, dairy cows, hogs, feedlot beef cattle, rangeland beef cattle, and lambs. When efficiency was measured in terms of the amount of fossil energy used to produce protein, the order was rearranged noticeably. The livestock that most efficiently converted fossil energy into protein were rangeland beef, followed by egg-laying hens, lambs, broiler chickens, hogs, dairy cows, and feedlot beef cattle.[10] Although all these studies used different bases of comparison, the conclusion in each case is substantially the same: more energy is expended in the production of livestock than is returned in edible food.

Is Feed Conversion Efficiency Declining?

A second criticism that has been voiced is that feed conversion efficiency is declining. For some time, there has been considerable debate over whether the current production methods have led to a decline in the conversion of feed into food. Reduced efficiency of feed utilization is significant, because it means that more pounds of grain are required to produce a pound of weight gain. Thus, even small declines in feed conversion effectiveness cause considerable consternation, since they are multiplied into large increases in the amount of energy consumption.

The recent improvements in animal management techniques have led to dramatic increases in production, but the changes in feed conversion efficiency have brought mixed results. One of the success stories of modern livestock production is in the broiler chicken industry, where the efficiency of feed utilization has increased about 47 percent since 1950.[11] The changes in feed conversion efficiency for cattle and hogs is less encouraging. Heichel and Frink report that the efficiency of hog production appears to have declined sharply about 1960 and then begun a slow improvement.[12] Based on U.S. Department of Agriculture data, they report a decline in the efficiency of feed utilization by cattle and calves of about 15 percent since 1950.[13] The Council for Agricultural Science and Technology disputes these conclusions and reports no such trend in either hogs or cattle.[14]

SUGGESTIONS FOR CHANGE

Agricultural researchers, farmers, and policy planners are intensely interested in discovering ways to increase the energy efficiency of the livestock rancher, poultry producer, and dairy farmer. This attention has resulted in numerous recommendations for change.[15]

Increased Grazing

Raising beef cattle can be accomplished by grazing animals entirely on rangeland, by feeding them in confinement feedlots, or by various methods that fall within these extremes. In general, a large percentage of beef cattle spend most of their lives grazing on rangeland and only the last five months in confinement feedlots being fattened for slaughter. In the aggregate, fattening cattle in confinement feedlots uses tremendous amounts of feed grains and nutritional supplements that embody significant quantities of nonrenewable energy. The resources devoted to producing feed grains—for example, land, irrigation water, fuels, machinery, and fertilizers—

could alternatively be used to produce crops for direct human consumption instead. Such a change would be likely to result in a greater yield of net energy, since crops grown for human consumption return more food energy per unit of energy input than do livestock.

Beef cattle and other ruminants are, of course, extremely useful food producers, because they have the ability to metabolize forage crops and food wastes that are nutritionally worthless to humans. It has been suggested that this ability to produce high quality food from land that is uncropable and from materials that humans cannot eat could be utilized by allowing cattle to graze on rangeland and crop wastes more extensively. If this were done, the energy cost of beef would be lower than under the current energy-intensive regime of feeding grains and feed supplements.

There are, of course, practical limitations to the suggestion to increase grazing. Agricultural policymakers and farmers must consider the economics of grazing versus confined feeding. Cattle are fattened on a diet of feed grains and supplements to help them mature into the better grades of beef, which are preferred by consumers and command higher prices on the market. Naturally, the cost of grain relative to the price of finished animals will affect the average time spent on the feedlot and the final slaughter weights, as producers respond to production costs and market prices. In addition, range production of beef can, under some conditions, require as much energy as feedlot beef, since cow ponies have been replaced by pickup trucks and cattle are driven to market by transport trucks.[16]

Recent changes in government beef grading standards and in retail marketing may accelerate the decision of cattle ranchers to graze cattle longer. The institutional changes occurred in 1976, when the U.S. Department of Agriculture revised the ways in which beef cattle are graded. Federally inspected beef cattle are graded in two ways, as to quality and yield. Quality grading, which is broken down into eight basic categories, measures the tenderness, juiciness, and flavor of the meat. Yield is a measure of leanness, indicating the percentage of the carcass useable as food after the waste fat has been trimmed off.

One major change was combining the two scales, so that federally graded beef is now identified for both yield and quality under all circumstances, which was not formerly done. The second important change was the decision to lower the minimum levels of marbling fat for the various quality grades of beef. Under the previous system of quality grading, it was assumed that an increasingly greater amount of marbling fat was needed as an animal grew older if its quality were to remain the same.

Both of these changes will affect energy consumption. The more important of the two is the revision in quality grading standards, which is expected to reduce by two weeks the average time an animal stays on the feedlot. This should result in a decrease of feed grain consumption of about 250 pounds an animal, reducing costs of production both in terms of energy and dollars. The second change—grading an animal as to both quality and yield—is likely to discourage ranchers from producing very fat cattle, whose production uses more feed and energy for each useable pound of meat.

Changes in retail marketing may also encourage cattle producers to reduce the time that cattle spend on energy-intensive feedlots. For example, one supermarket chain in the Washington, D.C., area has begun selling and promoting cuts of beef carved from cattle that are fed grain for only 100 days, about one-third less than the usual 150 days for conventionally raised beef cattle. The supermarket is actively promoting the meat to gain consumer acceptance, with the emphasis on lower price, possible health benefits, comparability of tenderness, and the savings in grain to help feed a hungry world.

Technological Innovations

Since Americans are likely to continue to eat energy-intensive feedlot beef, it has been suggested that the energy and dollar costs of producing that beef could be reduced by improving the efficiency of feed conversion. It is estimated that an improvement of only 10 percent would have decreased the feed grain requirements by more than 4 million metric tons in 1976.[17] For years, American cattle ranchers have attempted to keep down their costs of production and the prices of meat by using the hormone mimic diethylstilbestrol (DES). This drug increases weight gain for each pound of feed, but unfortunately, DES is generally considered carcinogenic. Its use has been banned in several foreign countries, and similar action seems likely in the United States. To replace DES, researchers have found new methods of improving feed conversion efficiency and increasing the energy efficiency of cattle production.[18]

Scientists have found that using the antibiotic monensin alters the metabolic products of microbes in the rumina of cattle and allows the cattle to get more energy from each unit of feed consumed. The antibiotic drug, which is marketed in the United States under the trade name Rumensin, can decrease feed consumption by more than 10 percent.[19]

Second, researchers have discovered a feed additive consisting of microencapsulated animal fats that increase the efficiency with which cattle convert roughages into weight gain, increases the portion of the carcass qualifying as USDA Choice, and boosts milk production in dairy cows. Using microencapsulated animal fats, a practice not yet approved by the Food and Drug Administration, can increase the efficiency of feed conversion and the average daily weight gain of cattle fed on roughages as much as 15 percent and can increase dairy cow milk production an average of 11 percent.[20]

The third breakthrough is the discovery of a plastic vaginal insert for heifers that increases the average daily weight gain and feed conversion efficiency. Feedlot heifers implanted with these devices show a 5 to 10 percent increase in average daily weight gain, and foraging heifers do even better, growing 13 to 22 percent faster. The net effect is to reduce the cost of feeding each heifer $15 to $20.[21]

An important question that should be considered by agricultural policymakers is whether the use of these devices and drugs, especially antibiotics, can adversely affect human health. So far, tests have indicated that these three methods of inducing weight gain are safe. If further study continues to show that these methods are safe, they can provide the means for increasing energy efficiency and decreasing energy use in the production of beef and milk.

Improved Operation and Maintenance

All livestock producers recognize that a successful, profitable farm or ranch depends on energy. One increasingly important use of energy is to heat and cool animal housing for greater production. Space heating and ventilation are essential in brooding poultry and finishing hogs. Supplemental heat and ventilation are also used, but to a lesser degree, in raising dairy cows, beef cattle, and sheep.

Using energy wisely to heat and cool livestock housing can reduce fuel requirements, improve the efficiency of feed conversion, and produce healthier livestock. Fortunately, all livestock producers can undertake many simple housekeeping maintenance procedures and simple changes in operation to save energy in heating and ventilating farm buildings. Basically, these changes boil down to planning carefully, conducting energy audits of heating and ventilation needs, insulating, routinely maintaining heaters and fans, and turning off pilot lights when not needed.[22]

Although energy can be saved by all livestock producers who adhere to a program of energy conservation, poultry producers have an unusual opportunity to save energy. Over 71 percent of the energy used in poultry production, or about 28 trillion BTUs, is consumed in brooding, so close management and maintenance could lead to substantial fuel savings.[23] These savings could total as much as 50 percent if poultry producers used partial house brooding, adhered to good management and maintenance practices, winterized the side curtains, and shut off the brooder pilot lights when they are not needed.[24] Adding insulation, selecting an energy-efficient ventilation system, using good ventilation practices, and sealing cracks around doors and windows are examples of other steps that can be taken to reduce energy consumption in poultry production.[25]

Some rather simple energy-saving hints and their potential for saving energy are shown in Table 5–1, which is reprinted from a study sponsored by the Electric Power Research Institute.

In addition to these changes in heating and ventilating livestock housing, other changes in maintenance and operation can significantly reduce energy consumption. All livestock producers can take care in their use of energy for general farm travel, lighting, feed processing and distribution, water supply, water heating, and waste handling.[26] Some of these suggested changes and their energy-saving potential are contained in Tables 5–2 and 5–3, which are reprinted from a study sponsored by the Electric Power Research Institute.

Table 5–1. Saving Energy in Farm Buildings

Broiler Houses

- Close up holes and cracks; insulate 10–20%
- Brood as near the center of the house as you can 5– 8%
- Brood as many chicks as the brooder can handle 4– 6%
- Adjust brooding temperature to conserve heat 3– 6%
- Keep litter dry 2– 4%

Livestock Buildings

- Cover windows with plastic if they are not essential
  for ventilation 10–12%
- Coordinate fan and heat thermostats 5–10%
- Clean and adjust thermostats 5– 8%
- Use correct ventilation 4– 6%
- Shutter fan outlets 2– 5%
- Clean heaters 2– 4%
- Use correct humidity 2– 3%

Estimated percentage savings based on total expenditure of energy use. Actual savings will vary depending on specific applications, but should fall in these ranges. These savings are not necessarily additive.

Source: G.B. Taylor, "Agricultural Energy Use," in *Efficient Electricity Use*, ed. Craig B. Smith, prepared for the Electric Power Research Institute (New York: Pergamon Press, 1976), p. 402.

Table 5-2. Saving Energy on Feedlots

• When hauling feed or cattle locally, match carrier to size of load and try to operate the vehicle with a full load . . .	15-20%
• Check personnel to be sure that they are making efficient use of energy-intensive and costly inputs such as feed, machinery and, of course, the cattle	10-20%
• Avoid placing cattle in excessively soft and muddy lots and provide shade in hot months, if possible. Construct concrete aprons in front of watering troughs	12-15%
• Find a market for manure to facilitate removal from feedlots. Meanwhile, stack in center of lot to serve as an elevated dry area. Also try to locate lot in crop production area to minimize hauling distances	10-12%
• Follow tips listed in crop production while operating heavy machinery	5-10%
• Take care in handling feed that none is lost	5-10%
• Keep dust, and hence incidence of infection, down by sprinkling with water during summer	5-10%
• Keep feed and silage dry and free from insects and rodents . .	6- 8%
• Be aware of local crop surpluses which can be used as supplementary feed	5- 8%
• Move cattle as few times as possible. The change in surroundings increases energy expenditure and usually takes weight off the cattle. Brand, inject, implant and dehorn all in one operation if possible	4- 8%
• Select the right type and size of motor for the job. A motor operating at half rated load may lose only 1 percent in efficiency but may lose 10 percent or more in power factor. This increases current and power losses over that of properly matched motors and loads	4- 6%
• When replacing machinery requiring fossil fuel for its operation, try to substitute diesel for gasoline engines	2- 5%
• Keep electric motors clean; clogged air ducts retard ventilation and may cause overheating and excessive energy use as well as possible need for replacement of the motor	2- 4%
• Shut off electric motors when equipment is not in use . . .	1- 2%

Estimated percentage savings based on total expenditure of energy use. Actual savings will vary depending on specific applications, but should fall in these ranges. These savings are not necessarily additive.

Source: G.B. Taylor, "Agricultural Energy Use," in *Efficient Electricity Use*, ed. Craig B. Smith, prepared for the Electric Power Research Institute (New York: Pergamon Press, 1976), p. 406.

Table 5–3. Saving Energy in Dairy Operations

• Insulate buildings that are cooled by refrigeration or heated	15–20%
• Consider using mercury vapor or high-pressure sodium vapor lamps instead of incandescent lights for outside lighting	8–10%
• Consider the feasibility of using recovery heat to supply or supplement hot water requirements	5–10%
• Use infrared heat rather than convected heat in barns where semi-open or open structures are used	5–10%
• Check animal water supplies and be certain that they are free of contamination and disease	4– 8%
• Use hot water wisely	4– 6%
• Make sure the milk refrigerator is operating properly	4– 6%
• Place lights properly so that they provide maximum illumination in areas where needed	4– 5%
• Provide adequate drainage to avoid mudholes in which the cattle can pick up diseases or decrease milk production . .	3– 5%
• Keep pipes carrying hot water well insulated	3– 5%
• Have available alternate energy sources such as a stationary engine to supply the required power and fuel if shortages occur	2– 5%
• Avoid excess washing of cows	3– 4%
• Plan to conserve on water use making maximum use of the water for cleaning the barns, etc.	2– 3%
• Consider the possibility of re-using wash water	2– 3%
• Set the hot water thermostat no higher than necessary . . .	2– 3%
• Put a timer switch on the pump control	1– 2%
• Repair leaks in water lines and faucets	1– 2%
• Reexamine water and power supply systems. Be sure that the pressure tank has an adequate amount of air. Turning the pump off and on often requires additional electrical energy	1– 2%

Estimated percentage savings based on total expenditure of energy use. Actual savings will vary depending on specific applications, but should fall in these ranges. These savings are not necessarily additive.

Source: G.B. Taylor, "Agricultural Energy Use," in *Efficient Electricity Use*, ed. Craig B. Smith, prepared for the Electric Power Research Institute (New York: Pergamon Press, 1976), p. 408.

Change Human Diets

Some agricultural commentators have suggested that energy could be conserved by reduced consumption of meat and animal products. The trend in this country and in other affluent countries is toward increased consumption of these foods. Reversing this trend would represent a significant sociological change and would not be accomplished easily. Such a change could, however, result in an enormous savings of fossil fuel and could release many acres of land for a less resource-intensive type of farming.[27]

Whether governments should become involved in creating policies that discourage people from eating animal products is a topic that will certainly stimulate lively debate. It seems appropriate that the individual should be the one to make the actual decision to change dietary habits, and the role of the state is best reserved to providing information and moral suasion (e.g., the "meatless Tuesdays" campaign of World War II).

STRATEGIES FOR CHANGE

States that wish both to increase farmers' profits and to conserve energy face a challenging task, yet the two goals need not be mutually exclusive. There are several strategies that states can devise to encourage all livestock producers to make changes that will improve energy efficiency and farm profits.

Provide Energy Conservation Information to Livestock Producers

Like all farmers, livestock producers understand that it is advantageous to keep production costs down. One way to reduce these costs is to eliminate any unnecessary use of fossil fuels and energy-intensive feed grains. States can help persuade livestock producers to curtail their use of these resources by providing them with information that shows the cost-effectiveness of particular energy conservation measures.[28] Education programs, energy audits, and demonstration projects are examples of how states can get the information to the farmer.

Initially, a state should examine whether the extension service at its land-grant college is properly funded and staffed. The agricultural extension service is probably the best way to provide livestock producers with the latest findings on energy-efficient, cost-effective production methods. Extension service personnel can keep farmers abreast of technological innovations and can offer advice on energy-efficient maintenance and operation procedures. A good outreach program might include classes, seminars, publications, and workshops.

Case studies that consider the economics of using energy-conserving methods are often very informative and persuasive. To take one example, large dairy farms require substantial amounts of fuel to provide hot water and supplemental space heating. One study of the economics of using a waste heat recovery system on a large dairy farm, performed by a natural gas company in Pennsylvania, reported favorable results.[29] Although the costs of installing a waste heat recovery system are about twice as great as for installing an electric heat system alone, the estimated savings will pay back the extra cost in less than two years. Over a ten-year period, the operating savings are expected to return the initial investment plus a return on investment in excess of 50 percent.[30] An investment like this is so economically beneficial that states should make an effort to inform farmers of the cost-effectiveness to overcome any initial hesitancy that farmers may feel.

To convince skeptics that investment in energy-effective equipment and methods is the right thing to do, states could fund projects that demonstrate the economic viability of energy conservation. It is, of course, the profit consideration that will influence livestock producers, and a demonstration project showing how a new method or feed supplement can enhance profits will be likely to win over the most ardent critics. The Small Farm Energy Project in Hartington, Nebraska, provides a good example of how the cost-effectiveness and viability of energy-saving methods can be demonstrated. Project participants recently completed construction of a solar collector for a dairy barn belonging to one of the project cooperators, an effort that enabled them to learn its operation and economics.[31]

Finally, states could offer advisory services such as energy audits to locate particular instances of energy waste. Trained experts would be able to detail specific examples of how farmers could tighten up their operations to reduce energy waste and increase profits. These audits could be conducted either by state agriculture personnel or by private consultants. States could, and probably should, charge farmers for some portion of the costs of the audit.

Provide Loan Programs to Help Farmers
Invest in Energy-Saving Methods

The operational changes discussed in the preceding section can help livestock producers reduce their direct consumption of fossil fuel energy. For example, equipment that uses solar energy has the potential for reducing the amount of fuel and energy devoted to heating and cooling livestock housing. In addition to these direct energy savings, the indirect consumption of energy can be reduced. For instance, beef cattle ranchers who use the new feed supplements and devices can improve the efficiency of feed utilization and reduce the amount of grain needed to produce a pound of beef, thereby saving the energy that went into the production of that grain.

Even though investment in solar devices or innovative equipment can reduce costs and energy use, livestock producers may hesitate to make the financial commitments necessary to reap these savings. This reaction is not uncommon, because some of the changes represent untested methods. States could encourage farmers to adopt innovative technologies, which have the potential to save large quantities of fuel and grain, by offering loans to farmers who will invest in these energy-conserving systems.[32]

A state loan program could take several forms, depending on the desired degree of state commitment. For example, a state could become an active participant, making direct loans with state funds at favorable interest rates. Alternatively, a state could help livestock producers obtain financing from traditional financial sources by guaranteeing private bankers that, in the event of default by the borrower, the state will repay the outstanding balance of the loan. States can make a loan guarantee program even more attractive by subsidizing the interest rate, giving farmers an incentive to invest and private bankers the assurance of a profitable loan.

Offer Tax Incentives to Livestock Producers

States can influence private investment decisions by offering livestock producers tax incentives for investment in technologically innovative energy-conserving equipment. The allowance of tax incentives—in the form of investment tax credits, tax exemptions, and tax deductions—can benefit both farmers and society. Farmers who invest in energy-efficient equipment that is eligible for tax preferential treatment save money, as the effective cost of investment is reduced. If farmers respond to the offer of tax incentives by investing in energy-saving equipment, society also benefits, as fuel is freed for other purposes.

Nevertheless, there are some competing considerations that should be recognized before a state enacts legislation affecting its tax structure. Unless a state reduces its expenditures to compensate for the revenue foregone by tax allowances, the revenue must be raised by shifting the tax burden to other taxpayers. This raises the issue of equity to those taxpayers who do not receive the benefits of tax preferences. Second, states must insure that the allowance of tax incentives does not create the opportunity for abuse. This could be done by limiting the availability of tax incentives only to investments that are both energy-efficient and economically risky or by setting strict limits on the maximum amount that can be claimed as a tax credit, deduction, or exemption.

CONCLUSION

From the foregoing discussion, it should be clear that producing livestock, particularly beef, requires more energy than most other farm commodities. Major reductions of energy use appear possible if farmers carefully audit their current use of energy, include energy conservation in any plans for purchasing new equipment or for upgrading their present buildings and machinery, properly insulate all livestock housing, and implement simple housekeeping changes in operation and maintenance.

Other possible changes have been suggested that are capable of having a far-reaching effect, but these major policy changes are probably beyond the realistic scope of state legislative action. Reduced feeding of grain to beef cattle and earlier slaughter would save energy, but would require dramatic modifications in production, marketing, and consumer acceptance. These changes are not impossible, but legislating such changes would touch on federally established standards and are probably best left to national energy policy.

In sum, states can actively encourage livestock producers to implement numerous energy-saving changes through the selective use of educational programs, demonstration projects, research and development grants, and financial incentives.

NOTES TO CHAPTER 5

1. *See* Earl O. Heady et al., *Agricultural and Water Policies and the Environment: An Analysis of National Alternatives in Natural Resource Use, Food Supply Capacity, and Environmental Quality*, CARD Report no. 40T (Ames: Center for Agricultural and Rural Development, Iowa State University, 1972); G.M. Ward, P.L. Knox, and B.W. Hobson, "Beef Production Options and Requirements for Fossil Fuel," *Science* 198 (October 21, 1977): 270.

2. G.H. Heichel and C.R. Frink, "Anticipating the Energy Needs of American Agriculture," *J. of Soil and Water Conservation* 30, no. 1 (January–February 1975): 49.

3. *Ibid. See also* Gary H. Heichel, *Auxiliary Energy Requirements and Food Energy Yields of Selected Food Crops*, Special Soils Bulletin 36 (New Haven: Connecticut Agricultural Experiment Station, 1974), p. 5.

4. William Lockeretz, *Agricultural Resources Consumed in Beef Production*, Report no. CBNS–AE–3 (St. Louis: Center for the Biology of Natural Systems, Washington University, 1975), p. 1.

5. U.S. Department of Agriculture, Economic Research Service, *National and State Livestock-Feed Relationships*, Statistical Bulletin no. 530 (Washington, D.C.: U.S. Government Printing Office, 1974).

6. Booz, Allen & Hamilton, Inc., *Energy Use in the Food System*, prepared for the Federal Energy Administration (Washington, D.C.: U.S. Government Printing Office, 1976), app. A(5).

7. *See* Roy N. Van Arsdall, *A Guide to Energy Savings for the Livestock Producer*, prepared for the U.S. Department of Agriculture and the Federal Energy Administration (Washington, D.C.: U.S. Department of Agriculture, 1977), p. 3; Gary G. Frank, *A Guide to Energy Savings for the Dairy Farmer*, prepared for the U.S. Department of Agriculture and the Federal Energy Administration (Washington, D.C.: U.S. Department of Agriculture, 1977), p. 3; Verel W. Benson, *A Guide to Energy Savings for the Poultry Producer*, prepared for the U.S. Department of Agriculture and the Federal Energy Administration (Washington, D.C.: U.S. Department of Agriculture, 1977), p. 3.

8. Heichel and Frink, *supra* note 2 at 50.

9. W.E. Splinter, "Alternative Energy Uses in Agriculture" (Lincoln: University of Nebraska, Department of Agricultural Engineering, n.d.), p. 4.

10. David Pimentel et al., "Energy and Land Constraints in Food Protein Production," *Science* 190 (November 21, 1975): 757.

11. Heichel and Frink, *supra* note 2 at 51.

12. *Ibid.*

13. *Ibid.*

14. Council for Agricultural Science and Technology, *Potential for Energy Conservation in Agricultural Production*, CAST Report no. 40 (Ames: Iowa State University, 1975), pp. 19–20.

15. The U.S. Department of Agriculture and the Federal Energy Administration have compiled three excellent energy conservation guides for livestock producers. *See* note 7, *supra*.

16. *See* Ward, Knox, and Hobson, *supra* note 1 at 265–71; Neil A. Patrick, "Energy Use Patterns for Agricultural Production in New Mexico," in *Agriculture and Energy*, ed. William Lockeretz (New York: Academic Press, 1977), pp. 36–37.

17. Thomas H. Maugh II, "The Fatted Calf: More Weight Gain with Less Feed," *Science* 191 (February 6, 1976): 454.

18. *Ibid.*, pp. 453–54. *See also* Thomas H. Maugh II, "The Fatted Calf (II): The Concrete Truth About Beef," *Science* 199 (January 27, 1978): 413.

19. Maugh, "The Fatted Calf: More Weight Gain with Less Feed," *supra* note 17 at 454.

20. *Ibid.*, pp. 453–54.

21. Maugh, "The Fatted Calf (II): The Concrete Truth About Beef," *supra* note 18 at 413.

22. *See* Van Arsdall, *supra* note 7 at 25–31; Frank, *supra* note 7 at 12–18; Benson, *supra* note 7 at 6–11, 17–21, 26–34.

23. *See* Benson, *supra* note 7 at 3–4, 6.

24. *Ibid.*, p. 6.

25. *Ibid.*, pp. 6–11, 17–21, 26–34. *See also* CAST Report no. 40, *supra* note 13 at 21; Georgia Department of Agriculture, *Farmers and Consumers Market Bulletin*, August 3, 1977, p. 8.

26. *See* Van Arsdall, *supra* note 7 at 13, 32–33; Frank, *supra* note 7 at 5–6, 19, 23, 26, 31–32, 36–38; Benson, *supra* note 7 at 12–13, 22.

27. Reducing the animal products portion of the American diet from 40 to 35 percent would result in an estimated saving of 134 trillion BTUs annually. *See* John D. Buffington and Jerrold H. Zar, "Realistic and Unrealistic Energy Conservation Potential in Agriculture," in *Agriculture and Energy*, ed. William Lockeretz (New York: Academic Press, 1977), p. 702.

28. The best examples of informative handbooks that show the cost-effectiveness of energy conservation measures are the U.S. Department of Agriculture and Federal Energy Administration guides to energy conservation for the livestock producer, the dairy farmer, and the poultry producer. *See* note 7, *supra.*

29. This case study was taken from G.B. Taylor, "Agricultural Energy Use," in *Efficient Electricity Use*, ed. Craig B. Smith, prepared for the Electric Power Research Institute (New York: Pergamon Press, 1976), pp. 415–20.

30. *Ibid.*, p. 417.

31. *Small Farm Energy Project Newsletter* no. 8 (Hartington, Nebraska: Small Farm Energy Project, Center for Rural Affairs, October 1977), pp. 1, 9. For more information on solar heating of dairy barns, *see* P. Thompson, *Solar Heating for Milking Parlors*, prepared for the U.S. Department of Agriculture, Agricultural Research Service, Farmers' Bulletin no. 2266 (Washington, D.C.: U.S. Department of Agriculture, 1977).

In addition, the Food and Agriculture Act of 1977 contains provisions for the establishment and operation of model farms and demonstration projects to promote the use of solar energy on farms. *See* 7 U.S.C.A. §§ 3261, 3262 (West Supp. February 1978).

32. State loan programs could follow the lead of the Food and Agriculture Act of 1977, which permits loans to be made for the purpose of "purchasing livestock, poultry, and farm equipment (including equipment which utilizes solar energy). . . ." 7 U.S.C.A. § 1942(a)(2) (West Supp. February 1978).

※ *Chapter 6*

Urban Agriculture

FARMING IN THE CITY:
POTENTIAL ENERGY SAVINGS

      Many U.S. urban centers and their surrounding suburbs are studded with small plots of land providing neither housing nor recreation, but merely lying vacant. Perhaps surprisingly, these scattered nonrural areas can be cultivated and planted to produce food and conserve energy. The idea that both city dwellers and suburbanites can produce some of their own food right in their own back yards is growing apace.

If the overriding concern of a state or locality is energy conservation, using vacant plots of land for food production is probably not the most energy-efficient idea. Rather, energy can be better conserved by erecting new housing on vacant land in urban areas and their close-lying suburbs. Construction of new homes on these parcels of land can help slow the spread of the suburban sprawl that compels many to live miles away from their jobs. Building homes close to work locations makes for an energy-efficient pattern of land use by reducing energy requirements for transportation and other activities.[1] Alternatively, vacant land could be transformed into recreational facilities to contribute to net energy savings. For people living in densely packed cities far from existing recreation areas, congenial recreation facilities nearer to home could mean substantial savings in energy, money, and time. Using vacant parcels of land to produce food ranks much lower on the list of energy-conserving activities. Nevertheless, urban agriculture can contribute to energy

153

savings. By encouraging people to plant gardens on publicly owned vacant land as well as on their own property, states and municipalities can help cut down energy consumption within the food system.

For the prospective gardener, energy conservation is probably not the primary consideration. A vegetable garden provides a source of recreation for the person who sits behind a desk five days a week. In addition, a vegetable garden is also thrifty. People on limited budgets may find the savings in food costs the most appealing aspect of their gardens. Energy conservation is a less tangible additional benefit.

Much has been written about the savings in energy and money that home gardening offers by authors running the gamut from the U.S. Department of Agriculture to the many so-called organic groups. Reports focus on every aspect of gardening from seed to harvest and reflect the diversity of attitudes among people who raise vegetable gardens.[2]

Home gardening can produce more food with less energy than commercial methods require, and these reports tell how. First, home gardeners can use the land more effectively than can commercial farmers. Tractors and other large farm machinery demand room to maneuver; the large turning areas left at the ends of fields cannot be planted. Most home gardeners, unencumbered by large machinery, can plant virtually all their land, including patches too small to be farmed commercially. Second, home gardening tends to be more skill- and labor-intensive and relies less on fossil fuel energy than does commercial agriculture. Some home gardeners save energy and money by substituting natural materials and methods for chemical fertilizers and pesticides. Often called "organic" farming, these methods substitute animal manure, composted wastes, and soil-enriching crops for commercially manufactured fertilizers, while natural predators and repellent crops replace synthetic pesticides to make gardens uninviting to harmful pests.[3]

Although success with these sophisticated techniques takes time and patience, the payoffs can be substantial. Ecology Action of the Midpeninsula, an environmental group in Palo Alto, California, has developed and published information on resource-conserving and energy-efficient home gardening methods. Reporting vegetable yields fourfold those of conventional farming methods with energy use down to a mere 1 percent of conventional levels, the California group found that use of chemical nitrogen fertilizer and irrigation water could be reduced markedly.[4] Depending on the condition of the soil, they applied at most 50 percent of the energy-intensive nitrogen fertilizer and irrigation water that conventional methods would require to produce the same amount of food on the same land.[5]

Besides conserving energy in the process of growing their food, home gardeners can circumvent the commercial food-handling and delivery chain and save additional energy. Eric Hirst estimates that the transporting, wholesaling, and retailing of food account for about one-fifth of the total energy expenditures in the food system.[6] Since home gardeners use their home-grown food directly, the energy that would have gone into transporting and distributing an equivalent amount of commercially grown food is saved. A second benefit is the elimination of the need for commercial packaging, an extremely voracious consumer of energy. Of course, any reduction of the glut of solid waste that chokes many cities today also improves the urban environment. Finally, the energy-intensive, refrigerated journey from distant field to dinner table is transformed into the short walk from the garden to the kitchen when commercial produce is replaced by home-grown. A home garden may not mean complete self-sufficiency, but it can mean energy saved and money in the bank.

The Garden of Eden had its serpent, and home gardens have their flaws. One problem that crops up in gardens located in urban and industrial areas is pollution. In many cities, such dangerous air pollutants as lead accumulate in the soil, so that vegetables grown in it may become contaminated. Some cities have had to scrape off some of the topsoil before crops could be grown. Another problem is that efforts to save energy by producing food at home may be thwarted in the food-processing stage. Most home canning methods consume far more energy than do large-scale commercial processing of the same amount of food. Home gardeners wishing to conserve must keep in mind that what counts is net energy consumption of all phases of food production—growing, harvesting, processing, and distributing—before the food is eaten.

In a time of runaway food bills, food additive scares, and growing discontent among the nation's farmers, it is perhaps not surprising that urban dwellers are bringing the country into the city. What is surprising is the extent to which many people are trying to gain some measure of control over their lives. The boom in urban agriculture has not been limited to vegetable gardening, though vegetable gardens have proliferated in the last decade. Most recently, it has taken the form of the urban homestead.[7] The vanguard of this movement is composed of suburbanites, who live normal suburban lives with one exception. Their early morning and evening hours are filled with tending to the chores necessary to raise chickens, ducks, sheep, goats, and even cattle in their own back yards.

The new interest in raising livestock in an urban environment has caught most city zoning and health officials and agricultural extension service county agents by surprise. Until local officials have time to take a fresh look at the zoning and health ordinances, the best advice for urban farmsteaders is to check the local ordinances and to confer with neighbors before doing anything.

IMPLEMENTATION STRATEGIES

Encourage Community Gardening by Making Government-Owned Land Available

A Gallup Poll released in January 1976 reported the growing popularity of vegetable gardening. More than half of those polled planned to grow vegetables that year, with 90 percent of the gardens to be located on homeowners' own property and the rest on community property or elsewhere.[8] Of survey respondents who did not plan on having vegetable gardens, 40 percent said they would if only land were available. More upper income households than lower income households had gardens, a finding that underscores a basic problem: many low income people are not reaping their share of the economic benefits that home gardening offers.

Publicly owned, reasonably accessible plots of land can accommodate potential gardeners looking for a place to dig. With enough land made available over a sufficiently wide area, people could cultivate their own gardens at a minimum of expense, travel, and inconvenience.

Several current municipal programs encourage the use of publicly owned lands to grow food. Boston's Project Revival is part of a neighborhood improvement program funded by the Community Planning and Development Program of the U.S. Department of Housing and Urban Development (HUD). The project has three land use components designed to encourage neighborhood food production: an ownership program, a permit program, and victory gardens.

Under the ownership program, the City of Boston sells municipally owned property to selected purchasers. Prior to sale, a piece of land is graded to its natural contour and blanketed with enough topsoil to support a garden. During the project's first nine months of operation, one hundred lots were sold at unusually low prices ($100 and $200 a lot) and fifty more readied for sale. To ensure that these lots are used for food production and for no other purpose, Boston placed restrictions on the sales. The minimum lot size required for a building permit in Boston is 5,000 square feet, and the city has limited the size of the garden plots, accordingly, to less than 5,000 square feet. In addition, most lots are sold only to homeowners living adjacent to them, although the city has sold some land to absentee landlords wishing to rent garden space to their tenants.

Outright sale of municipally owned land, particularly at such low prices, may not be a good idea for most cities. For a city to relinquish the right to use the land differently in the future, perhaps for multifamily housing or public recreational facilities, seems unwise. Alternatively, lots might be leased under use permits so that land use options remain open for the future.

The other two programs of Project Revival provide land to gardeners on a temporary basis. The permit program allows applicants to garden on city-owned property for a year at a time. Of the three garden programs, the city spends the least money on the permit program, and land preparation is not up to the standards of the other programs.

The third component of Project Revival, the group garden program, is named for the World War II victory gardens. Overseen by the City Parks and Recreation Department, these pieces of land are loaned on a temporary basis to groups of citizens. Groups applying for membership in the program must submit a plan for the management of the garden before a permit is granted. The group garden program is designed specifically for people with low incomes, the elderly, and children. Because of this focus on the disadvantaged, the group garden project is eligible for federal funds from the HUD Community Planning and Development Program.

On spring days in Washington, D.C., vintage World War II victory garden plots are tended devotedly just as they have been for the last thirty-five years. Interest in the gardens, located on federally owned land, is so great that people who signed up years ago are still waiting for a space. In recent years, the National Parks Service has prepared additional federally owned property for gardening under a program called The Green Scene. Although this program has recently been phased out, other programs have expanded to help fill the gap. The Cooperative Extension Service administers community garden programs on city-owned land and since 1975 has spent considerable effort to make garden plots available to inner city residents. The combined work of the National Parks Service, the Cooperative Extension Service, and the District of Columbia Recreation Department has provided hundreds of gardening sites and has given Washington residents one of the largest community gardening programs in the nation.

A city or state can convert vacant lots into community gardens with relative ease and at little expense. Overhead costs for a program granting temporary permits are likely to be minimal, and gardeners can be charged a small fee to cover these expenses. Garden projects funded by the HUD Community Planning and Development Program, like Boston's victory gardens, customarily provide land and irrigation water at no charge. Other programs, such as the Washington, D.C., victory gardens, charge gardeners $10 a year for a plot of 1,250 square feet. The fee covers local administration by a gardening organization, use of the land, and irrigation water. Even such small fees could be waived for needy people.

Some people may hesitate to invest time and effort in land that may be taken from them when their annual permits run out. Ecology Action of the Midpeninsula and other groups have found that home garden productivity rises as the soil is conditioned over a period of several years. It may take two or more years of cultivation to make soil easy to work and fully productive, with food yield high and energy consumption low. If gardeners had some assurance that they could farm the same plot from one year to the next, they might be more willing to work diligently at improving the soil. Preferential allocation of permits could provide that assurance.

Encourage Community Gardening by Providing Preassembled Home Gardening Kits

Some people may not know how to go about planting a vegetable garden. Everyone wishing to raise a garden, whether on his own property or on community land, should be encouraged to do so through dissemination of gardening information. U.S. Department of Agriculture publications are full of information invaluable to people wishing to grow their own food. Most USDA publications are free for the asking; states and cities could let potential gardeners know that these and other publications are available.[9]

For people who want to avoid chemical farming methods, a number of pamphlets explain how to raise vegetables organically. A useful home gardening kit might consist of a booklet outlining how to raise a vegetable garden, packets of seeds likely to grow well in the particular area, fertilizer, and a selected reference list of additional information sources such as the state agriculture department, agricultural extension services, and gardening books and periodicals. States and municipalities can buy seeds and fertilizer in large quantities and sell them cheaply, thus enhancing the economic appeal of home gardening.

Most recipients of these gardening materials should probably pay a fee to help defray the costs of the program (with exceptions for the needy) for two reasons. First, not all taxpayers will benefit from the program; those who do should bear most of the cost. Second, most people place a greater value on something that costs money than on something that comes free.

CONCLUSION

At the outset, it must be conceded that most suburbanites and city dwellers are interested in urban agriculture for reasons other than energy conservation. Farming in the city offers an opportunity to achieve a degree of self-sufficiency, to obtain tasty garden-fresh food, to save some money, and to relax in a healthy leisure activity. Producing food in our cities is unlikely to reduce overall energy consumption dramatically, but it does provide one way for many people to conserve small amounts of energy while they receive the other, more tangible benefits.

NOTES TO CHAPTER 6

1. A companion book in this series details many land use strategies that can be used to conserve energy. *See* Corbin Crews Harwood, *Using Land to Save Energy* (Cambridge, Massachusetts: Ballinger Publishing Co., 1977).

2. *See, e.g.,* U.S. Department of Agriculture, *Gardening for Food and Fun: Yearbook of Agriculture* (Washington, D.C.: U.S. Government Printing Office, 1977); Robert Rodale, ed., *The Basic Book of Organic Gardening* (New York: Ballantine Books, 1975).

3. Farmers and home gardeners should soon have a better idea of the practicality, desirability, and feasibility of collecting and spreading organic waste materials on their land. Pursuant to the Food and Agriculture Act of 1977, the secretary of agriculture is directed to conduct and submit a report on these issues by September 1978. *See* 7 U.S.C.A. § 3303 (West Supp. February 1978).

4. Ecology Action of the Midpeninsula, "Resource-Conserving Agricultural Method Promises High Yields" (Palo Alto, California: Ecology Action of the Midpeninsula, 1976), p. 1. Contact:

> Ecology Action of the Midpeninsula
> 2225 El Camino Real
> Palo Alto, CA 94306
> (415) 328–6752

5. Irrigation water and nitrogen fertilizer use ranged from approximately 6 to 50 percent of that used by standard farming practices. *Ibid.*, pp. 7–8.

6. Eric Hirst, "Food-Related Energy Requirements," *Science* 184 (April 12, 1974): 135.

7. For an excellent description of this growing phenomenon, with examples of how some people in the Washington, D.C., area are beginning to work toward self-sufficiency, *see* Adrienne Cook, "The Urban Farmstead," *Washington Post*, December 11, 1977, Magazine Section, pp. 24–32.

8. This Gallup Poll was conducted on behalf of Gardens for All, Inc., a non-profit organization located in Shelburne, Vermont.

9. The Cooperative Extension Service at the University of the District of Columbia uses a general mailing to residents of Washington, D.C., telling where gardening information is available. In addition, a telephone hotline is maintained for those who want an immediate response to an agricultural problem.

Conclusion

In the introduction, this book identified as its purpose a description of barriers to energy conservation, a presentation of suggested energy-conserving techniques, and an analysis of various energy-saving implementation strategies. Prior to the oil price hikes of the last four years, the national priority accorded energy conservation was very low. In agriculture, a farmer's production costs attributable to energy use were relatively small, and thus, understandably, agricultural research did not concentrate on energy conservation. The Arab oil embargo of 1973–1974 jolted all Americans, causing policymakers to reconsider our national energy policy and researchers to hurriedly seek alternatives. In agriculture, these reports catalogue how energy is consumed in farm operations and how it can be saved. The results of this research point to one conclusion: farmers can economically reduce their energy consumption and increase their energy efficiency using existing technology.

The preceding chapters have carefully refrained from suggesting one "best" approach to enhance energy conservation on farms. Rather, a selection of different energy-saving techniques and implementation strategies has been presented for several reasons. Agricultural research is an ongoing pursuit that may lead to a reevaluation of current techniques or the development of entirely new ideas. In addition, farming operations and energy use vary from state to state and from farm to farm. Thus, no single set of energy conservation strategies is universally applicable to accommodate this mix of agricultural commodities and production methods. Finally, differences in geography, climate, politics, economics, and social conditions require policymakers to make appropriate modifications in suggested strategies to maximize social welfare.

Most observers agree that what is needed is an integrated, coordinated agricultural energy policy to accommodate the nation's economic, environmental, social, and political needs. Developing a sound approach to energy conservation invites the close partnership of the federal government, states, localities, and individual farmers. State and local energy conservation programs can provide the fine tuning needed to complement an overall national farm policy that seeks to maximize farm output and profits, husband our natural resources, maintain the fertility of the soil, and hold the line against food price increases.

The potential for increased energy efficiency on farms is vast, like the hidden fertility of the unplowed Great Plains a century ago. Realizing that potential is an opportunity too momentous to be ignored.

Index

About the Author

Robert A. Friedrich, staff attorney at the Environmental Law Institute, is a graduate of the University of California, Los Angeles, and Columbia Law School. Before joining ELI in 1975, he taught at the University of Wisconsin Law School and was a staff attorney for the Judicial Panel on Multidistrict Litigation. In addition to his work on the Energy Conservation Project, Mr. Friedrich is co-author of a report on legal barriers to solar heating and cooling of buildings and is currently helping prepare a report on legal guarantees of access to sunlight for solar energy systems.

BIBLIOTHEQUE VANIER | VANIER LIBRARY
UNIVERSITÉ